# Pearson Education
# Test Prep Series for

# AP® Environmental Science

## Myra Morgan

**To accompany:**

# ENVIRONMENT:
## The Science behind the Stories
AP® Edition
Sixth Edition

## Jay Withgott • Matthew Laposata

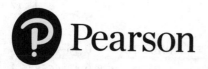 Pearson

330 Hudson Street, NY, NY 10013

Executive Editor: Alison Rodal

Courseware Director, Content Development: Ginnie Simione Jutson

Managing Producer, Science: Michael Early

Content Producer, Science: Margaret Young

Production Management and Composition: Norine Strang, Cenveo Publisher Services

Rights & Permissions Manager: Ben Ferrini

Manufacturing Buyer: Maura Zaldivar-Garcia, LSC Communications

Product Marketing Manager: Christa Pesek Palaez

Cover Photo Credit: Chris Cheadle/Alamy Stock Photo

1  17

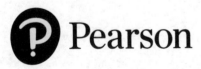 Pearson

ISBN-10:     0-13-465826-4

ISBN-13: 978-0-13-465826-1

www.PearsonSchool.com/Advanced

# Brief Contents

# About the Author

*Myra Morgan* has been actively involved in teaching science for the past 30 years as Head of Sciences at Miss Porter's School (CT), The Ethel Walker School (CT), and the Sage Hill School (Newport Coast, CA). As part of her teaching in AP Environmental Science, AP Biology, and many field research courses over the years, Myra has given students opportunities to study with her at the Bermuda Biological Station, at the National Undersea Research Center, and in the Gulf of Maine using ROV and submersible technology. Myra is an AP Environmental Science College Board Consultant and has been involved as an AP Reader beginning with the first year of the AP Environmental Science course in 1998. She provides AP Summer Institutes and workshops throughout the country in both biology and environmental science and has served as an AP Mentor to many teacher cadres in the College Board's New England and Middle States Regions. Myra has worked for the National Math & Science Initiatives in Kentucky, Connecticut, and Massachusetts, and currently is the AP Environmental Science College Board consultant for MASS Insight Education (Boston, MA). She has been honored for her contributions to the sciences and the AP Program with an Advanced Placement Recognition Award from the College Board, the NABT Outstanding Biology Teacher Award, a RadioShack Science Teaching Excellence Award, the Teaching Chair of Distinction in Science (Miss Porter's School), and a Teacher Fellowship at the University of St. Andrews, Scotland.

# Part I

# Introduction to the AP® Environmental Science Course and Exam

Welcome to the exciting and rewarding journey that is the AP Environmental Science course! Of all of the AP science courses, AP Environmental Science is unique in that it focuses on the relevant and pressing problems that face our species and the planet. It is an applied science that focuses on problem solving and on understanding the nature of environmental issues by emphasizing *critical thinking* and encouraging you to immerse yourself in the *science* behind the issues. Laboratory and field investigations also form an essential component of the course, as these experiences afford critical insights into how science is practiced, how environmental problems are studied, and how humans worldwide are affecting their environment.

Part I of this Test Prep Series for AP gives an overview of the Advanced Placement® program and the AP Environmental Science Examination®. It also describes the nature of the AP Environmental Science course, introduces some specifics as to the types of questions encountered on the exam, and provides helpful test-taking strategies. Finally, a correlation chart is included that shows where in the text, *Environment: The Science behind the Stories*, Sixth Edition, by Withgott and Laposata, you will find key information that commonly appears on the AP Environmental Science Examination.

Part II of this Test Prep Series for AP is keyed to *Environment: The Science behind the Stories*, Sixth Edition, by Withgott and Laposata, and provides a focused review of the major environmental science topics. It gives an overview of important information found in each text chapter and also provides sample multiple-choice and free-response questions. These chapter summaries are not a substitute for the text, class discussions, or laboratory portion of the course; however, they do offer a review that will help focus your preparation for the exam.

Each Part II chapter summary is organized according to *"You Must Know"* and should be reviewed and studied before you attempt the practice multiple-choice and free-response questions provided. Finally, be sure to thoroughly review the answers and explanations to prepare yourself for the range of questions and depth of information you will encounter on the AP Environmental Science Examination. This thorough approach will support a solid understanding and a confident ability to *apply and connect* the major concepts in the AP Environmental Science course.

Part III contains all of the answers and explanations for the Part II multiple-choice and free-response questions. Finally, Part IV includes a practice test with answers and explanations.

## *AP Environmental Science: A Brief Course Description*

AP® Environmental Science is one of the most relevant and exciting Advanced Placement courses offered to qualified high school students today. This course brings to life many topics in the news and creates an understanding of the science behind the many diverse environmental issues affecting all of our lives. Because of its interdisciplinary nature, AP Environmental Science is an excellent option for any student who has completed two years of a high school laboratory science: one year of life science and one year of physical science, often biology and chemistry. Because the course requires quantitative analysis, students should also have taken at least one year of algebra. Because of these prerequisites, AP Environmental Science will usually be taken in either the junior or senior year of high school.

The AP® Program is sponsored by the College Board, a nonprofit organization that oversees college admissions examinations. The College Board summarizes the goals of the AP Environmental Science course: "to provide students with the scientific principles, concepts, and methodologies required to understand the inter-relationships of the natural world, to identify and analyze environmental problems both natural and human-made, to evaluate the relative risks associated with these problems, and to examine alternative solutions for resolving or preventing them."

The interdisciplinary nature of AP Environmental Science embraces a wide variety of topics. Yet there are several major unifying themes that cut across the many topics included in the College Board's topic outline. The College Board lists six themes that should be stressed in all AP Environmental Science courses. These themes are from pages 4–5, *AP Environmental Science Course Description*, published by the College Board.

- Science is a process.
- Energy conversions underlie all ecological processes.
- The Earth itself is one interconnected system.
- Humans alter natural systems.
- Environmental problems have a cultural and social context.
- Human survival depends on developing practices that will achieve sustainable systems.

The following College Board abbreviated topic outline describes the scope of the AP Environmental Science course and exam. The percentage after each major topic reflects the approximate proportion of multiple-choice questions on the exam that pertain to that heading; thus, the percentage also indicates the relative emphasis that should be placed on these topics in order to best prepare for the AP® Exam.

**Source: Copyright © 2016 The College Board. Reproduced with permission. http://apcentral.collegeboard.com.**

**I. Earth Systems and Resources (10–15%)**
   A. Earth Science Concepts
   B. The Atmosphere
   C. Global Water Resources and Use
   D. Soil and Soil Dynamics

**II. The Living World (10–15%)**
   A. Ecosystem Structure
   B. Energy Flow
   C. Ecosystem Diversity
   D. Natural Ecosystem Change
   E. Natural Biogeochemical Cycles

**III. Population (10–15%)**
    A. Population Biology Concepts
    B. Human Population

**IV. Land and Water Use (10–15%)**
    A. Agriculture
    B. Forestry
    C. Rangelands
    D. Other Land Use
    E. Mining
    F. Fishing
    G. Global Economics

**V. Energy Resources and Consumption (10–15%)**
    A. Energy Concepts
    B. Energy Consumption
    C. Fossil Fuel Resources and Use
    D. Nuclear Energy
    E. Hydroelectric Power
    F. Energy Conservation
    G. Renewable Energy

**VI. Pollution (25–30%)**
    A. Pollution Types
        1. Air pollution
        2. Noise pollution
        3. Water pollution
        4. Solid waste
    B. Impacts on the Environment and Human Health
        1. Hazards to human health
        2. Hazardous chemicals in the environment
    C. Economic Impacts

**VII. Global Change (10–15%)**
    A. Stratospheric Ozone
    B. Global Warming
    C. Loss of Biodiversity
        1. Habitat loss; overuse; pollution; introduced species; endangered and extinct species
        2. Maintenance through conservation
        3. Relevant laws and treaties

# The AP Environmental Science Exam

The AP Environmental Exam is three hours in length and consists of two sections: a 90-minute multiple-choice section and a 90-minute free-response (essay) section. The exam is challenging and requires the ability not only to understand the science behind the environmental issues, but also to make important connections between those issues. For example, what do ocean acidification, tropical rainforest destruction, and using fossil fuels as an energy source have in common? Understanding the actual *evidence* for global climate change, being able to *analyze and interpret* scientific data tables, and then to offer *concrete approaches or solutions* to the many environmental issues discussed in the course are all assessed in the exam. You should be prepared to recall basic environmental facts and concepts, to apply these to the analysis of particular environmental problems, to synthesize facts and concepts making essential connections between environmental topics, to perform quantitative analysis, and to demonstrate reasoning and analytical skills.

# Section I of Exam: Multiple-Choice Questions

Section I of the AP Environmental Science Exam contains 100 multiple-choice questions that test both scientific facts and their application and will constitute 60% of your final grade. You will have 90 minutes to complete these 100 questions, so you should spend an average time of less than a minute on each question. *Multiple-choice scores are based only on the number of questions answered correctly.* Points are no longer deducted for incorrect answers so that students are encouraged to answer all questions. The AP Environmental Science Exam will contain several different types of multiple-choice questions, each of which are listed and described below, with an example.

## Traditional Questions

The first types are the traditional multiple-choice questions, requiring basic factual recall. These will be straightforward questions with five options and one correct answer. Here is an example.

1. The level of ecological organization described by field ecologists when studying multiple interacting species that live in the same area is
   (A) the habitat.
   (B) the energy flow.
   (C) the population.
   (D) the community.
   (E) the ecosystem.

   **Answer: D.** A community is made up of numerous populations of different species in a given area. In contrast, the ecosystem level would include the community of different populations *plus* the specific environmental factors (temperature, soil, moisture, light, etc.) in that area.

## "Reverse" Multiple-Choice Questions

For some students, this is a slightly more difficult type of question because it features four answers that are correct and only one that is incorrect. You are asked to determine the incorrect choice. Generally, these questions contain the word *not* or the word *except*. Here is an example.

2. Which of the following is *not* a major greenhouse gas?
   (A) carbon dioxide
   (B) nitrous oxide
   (C) chlorofluorocarbons
   (D) water vapor
   (E) carbon monoxide

   **Answer: E.** Water vapor, perhaps surprisingly, is the most common greenhouse gas, since it absorbs more infrared radiation from Earth than any other compound. Other greenhouse gases include $CO_2$, $N_2O$ (nitrous oxide), and CFCs (chlorofluorocarbons). Carbon monoxide (CO), although very toxic, is *not* a major greenhouse gas.

## Least and Most Likely Multiple-Choice Questions

This type of question often has two or more choices that are correct, but only one that is the *least* or *most* likely. It is essential that you read *all* options carefully, never deciding on the correct answer until you have ranked each choice carefully. Here is an example.

3. When the IPAT model is applied to the United States, the *least* important factor in determining the environmental effects of the U.S. population is
   (A) the value of P.
   (B) the value of A.
   (C) the value of T.
   (D) the value of A and T.
   (E) the value of TFR.

   **Answer: A.** The size of the U.S. population (P) is fairly insignificant compared to the great importance of affluence (A). Affluence is directly related to the consumption of Earth's natural resources, and it has been determined that environmental impact is not so much about the numbers of individuals, as about what those individuals are doing. The addition of 1 American has as much environmental impact as the addition of 4.5 citizens of China, 10 citizens of India, or 19 citizens of Afghanistan, all because individuals from affluent societies leave a considerably larger per capita ecological footprint.

### *Roman Numeral Questions*

There are usually only a few Roman numeral questions. With this type of question, three or four answers are labeled with Roman numerals, and one or more of the answers given could be correct. You are then provided with a series of options from which to select the correct Roman numeral or set of Roman numerals. For example:

4. Catalytic converters remove which of the following from automobile exhaust?

      I. Carbon monoxide    II. Particulates   III. Nitrogen oxide

(A) I only
(B) II only
(C) III only
(D) I and II only
(E) I and III only

**Answer: E.** Catalytic converters remove carbon monoxide from exhaust by oxidizing it to carbon dioxide. Also, nitrogen oxides are reduced to nitrogen and oxygen; however, particulates are *not* removed by catalytic converters.

### *Data Analysis and Interpretation Questions*

Data analysis and interpretation are skills at the heart of doing science; therefore, this type of question occurs frequently. You will be presented with a set of data in a graph, table, or chart format, requiring your analysis and interpretation of the meaning of the data. Often there is more than one question that makes use of the same data. These questions may also require that you do some calculations to determine the correct answer. *Calculators are not permitted on the AP Environmental Science exam*, so you should complete all required work in the test booklet before making your choice.

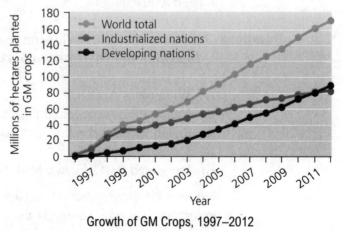

Growth of GM Crops, 1997–2012

5. Examining the graph between 2005–2009, there is evidence that
    (A) the rate of GM crop growth was being controlled.
    (B) the rate of increase of GM crops in industrialized nations was greater than in developing nations.

(C) the rate of increase of GM crops in developing nations was greater than in industrialized nations.

(D) the rate of growth of GM crops in both developing and industrialized nations was equal.

(E) not all nations are being accounted for.

6. After examining the graph to determine the percent change of the world's GM crops between 2008–2011, one concludes that GM crops increased by approximately ___ during this time.

(A) 160%

(B) 80%

(C) 40%

(D) 33%

(E) 15%

**Answers: (#5) C.** The question is asking for a comparison of *rates* of increase between industrialized and developing nations, not totals. The slopes of the two graph lines need to be compared between 2005–2009, in order to determine the rate of increase. One can visually see that the developing nations' graph line is steeper than that of the industrialized nation during this time, indicating a faster rate of growth. **(#6) D.** In order to determine the world's percent change in growth of GM crops between 2008–2011, one uses the formula:

$$\% \text{ Change} = (\text{Final} - \text{Initial}) \div (\text{Initial}) \times 100$$
$$\% \text{ Change} = (160 - 120) \div (120) \times 100 = \mathbf{33.3\%}$$

## Classification Questions

This type of question consists of 5 lettered choices followed by a list of statements. You are asked to select a lettered choice that best fits each statement. In each set, a choice may be used once, more than once, or not at all. Here is an example.

**Questions 7–9 refer to the following environmental list of U.S. and international environmental regulations and treaties.**

(A) Wilderness Act

(B) National Environmental Policy Act

(C) Resource Conservation and Recovery Act

(D) Montreal Protocol

(E) Integrated Pest Management Act

7. Requires an environmental impact statement for any major federal project, such as one that might use pesticides near a wilderness area.

8. Focuses on avoiding ozone depletion of the stratosphere.

9. Contains an amendment that specifies how to manage sanitary landfills to avoid environmental contamination.

all questions. The free-response section counts for 40% of your examination grade. Each of the four essays is weighted equally, with a 10-point potential for each essay response. The raw scores of the AP Environmental Science Examination are converted to the following 5-point scale.

> 5 – Extremely Well Qualified
> 4 – Well Qualified
> 3 – Qualified
> 2 – Possibly Qualified
> 1 – No Recommendation

Some colleges give undergraduate course credit to students who achieve scores of 3 or better on AP exams. Other colleges require students to achieve scores of 4 or 5. Each college and university sets its own AP credit and placement policies, and because these policies are individual and variable, it is important that you check the specific policies of the college or university for which you are interested on the College Board website (www.collegeboard.com/ap/creditpolicy).

# Test-Taking Strategies for the AP® Environmental Science Free-Response Questions

- You will have 90 minutes to answer the four free-response questions. Plan on approximately 22 minutes per essay. Since each essay is of equal value (10 points), dividing your writing time equally is important.
- **Read all questions** thoroughly before beginning to write. Determine which question seems to be the one about which you are most confident. *Answer this question first*, so that IF you run short of time on any question, it will be your most challenging one.
- **Write an essay!** Always answer in complete sentences. Outlines are unacceptable, and bulleted lists are not awarded any points and will not be graded. Diagrams will receive credit only if they are relevant, well-labeled, and are connected to the text of your essay.
- **Read questions carefully.** Check to see whether there are comparisons to be made, an experiment to design, an *economic* advantage to describe, an *environmental* benefit to be explained, or data to be analyzed. Then do just that. Write a response that specifically answers the question that has been asked. Go into detail, defining terms as long as they are relevant to the subject and to the point. You will earn points only for sentences that *specifically* answer the question or provide illustrative examples.

▌ **Never make general statements.** For example: activity "x" causes "pollution," activity "y" is "bad for the environment." These statements are much too vague and will receive NO credit. Follow through completely with your thoughts or examples, providing a full explanation for any term you use.

▌ **Label your answers** so there is no doubt where your answer is. All essays have very specific parts, such as *a, b, c,* or *i, ii, iii.* This also helps ensure that you have finished answering the question and can move forward to the next question.

▌ **Do not restate the question** in your answer as it is a waste of your time. You also do not need to have an introduction or conclusion to your essay. Science writing is specific and concise.

▌ **If given a choice of topics to write about, make your choice**(s) and stick with it (them). If two examples are asked for, give *only* two. No additional examples will be graded, and you earn no "extra credit" by providing more examples.

▌ **Write clearly and neatly** using a ballpoint pen with black or dark blue ink. Neatly cross out mistakes to save time, rather than erasing or using white-out. If the grader cannot read what you wrote, you will receive no credit.

▌ Try not to panic if you think you are somewhat unfamiliar with a question's topic. You will probably know something about the topic, so write what you know. **Do not give up and leave the entire question blank.**

▌ **Routinely practice your math!** Math-based free-response questions can be challenging. However, if you practice these types of essays throughout the year, you will gain confidence and become more successful. Remember to **show your work** and **label all units** to receive credit on the math questions. Every AP Environmental Science student should be comfortable working with percent change, metric prefixes, basic algebra, scientific notation and exponents, and dimensional analysis.

# AP Topic Correlation for *Environment: The Science behind the Stories*, Sixth Edition, by Withgott and Laposata

The following chart is intended to help you study for the AP® Environmental Science Exam. The left column includes the AP Environmental Science topics as they appear in the College Board topic outline. The right column includes the corresponding chapters and topics in your textbook, *Environment: The Science behind the Stories*, Sixth Edition, by Withgott and Laposata. One can easily observe that some topics are found in several chapters, emphasizing the *interconnectedness* of topics. Use this outline to keep track of your review, making sure to cover every topic. Check off each topic when you have thoroughly reviewed your text and the focused chapter review provided by this Pearson Education Test Series.

| CORRELATED TO THE COLLEGE BOARD TOPICS FOR AP® ENVIRONMENTAL SCIENCE | CORRELATION TO ENVIRONMENT: THE SCIENCE BEHIND THE STORIES, WITHGOTT & LAPOSATA, 6E |
|---|---|
| **I. Earth Systems and Resources (10–15%)** | **Chapters 2, 9, 15–18** |
| A. Earth Science Concepts (Geologic time scale; plate tectonics, earthquakes, volcanism; solar intensity; and latitude) | 2, 17 |
| B. The Atmosphere (Composition; structure; weather and climate; atmospheric circulation; Coriolis Effect; atmosphere–ocean interactions; ENSO) | 16–18 |
| C. Global Water Resources and Use (Freshwater/saltwater; ocean circulation; agricultural, industrial, and domestic use; surface and groundwater issues; and conservation) | 15–16 |
| D. Soil and Soil Dynamics (Rock cycle; formation; composition; physical and chemical properties; main soil types; erosion and other soil problems; soil conservation) | 2, 9 |

## II. The Living World (10–15%)                    Chapters 2–6, 11, 18

A. Ecosystem Structure                                          3, 4, 5
(Biological populations and communities;
ecological niches; interactions among species; species diversity and edge effects;
major terrestrial and keystone species; aquatic biomes)

B. Energy Flow                                                  2, 4, 5
(Photosynthesis and cellular respiration; food
webs and trophic levels; ecological pyramids)

C. Ecosystem Diversity                                          3–6, 11
(Biodiversity; natural selection; evolution;
ecosystem services)

D. Natural Ecosystem Change                                     4, 18
(Climate shifts; species movement; ecologi-
cal succession)

E. Natural Biogeochemical Cycles                               2, 5
(Carbon, nitrogen, phosphorus, sulfur, water,
conservation of matter)

## III. Population (10–15%)                         Chapters 1, 3, 8, 10, 11

A. Population Biology Concepts                                  3
(Population ecology; carrying capacity;
reproductive strategies; survivorship)

B. Human Population                                             1, 8
1. Human population dynamics
(Historical population sizes; distribution;
fertility rates; growth rates; doubling
times; demographic transition;
age-structure diagrams)

2. Population size                                           8
(Strategies for sustainability; case
studies; national policies)

3. Impacts of population growth                              8, 10, 11
(Hunger; disease; economic effects;
resource use; habitat destruction)

## IV. Land and Water Use (10–15%)          Chapters 3–4, 6–7, 9–16, 19, 22–24

A. Agriculture
1. Feeding a growing population                              9, 10, 12

(Human nutritional requirements; types
of agriculture; Green Revolution; genetic
engineering and crop production; deforestation; irrigation; sustainable agriculture)

2. Controlling pests                                                      10, 14
   (Types of pesticides; costs/benefits
   of pesticide use; integrated pest
   management; relevant laws)

B. Forestry                                                              12
   (Tree plantations; old growth forests; forest
   fires; forest management; national forests)

C. Rangelands                                                           9
   (Overgrazing; deforestation; desertification;
   rangeland management)

D. Other Land Use
   1. Urban land development                                            13
      (Planned development; suburban sprawl;
      urbanization)
   2. Transportation infrastructure                                     12, 13, 15
      (Federal highway system; canals and
      channels; roadless areas; ecosystem
      impacts)
   3. Public and federal lands                                          3, 6, 11, 12, 19
      (Management; wilderness areas; national
      parks; wildlife refuges; forests; wetlands)
   4. Land conservation options                                         3, 4, 6, 11, 12, 22, 23
      (Preservation; remediation; mitigation;
      restoration)
   5. Sustainable land-use strategies                                   9, 11–13

E. Mining                                                               6, 19, 23
   (Mineral formation; extraction; global
   reserves; relevant laws and treaties)

F. Fishing                                                              10, 16
   (Fishing techniques; overfishing;
   aquaculture; relevant laws and treaties)

G. Global Economics                                                     6, 7, 24
   (Globalization; World Bank; Tragedy of the
   Commons; relevant laws and treaties)

**V. Energy Resources and Consumption (10–15%)**      **Chapters 1–2, 8, 13, 15, 18–21**

A. Energy Concepts                                                      2, 19
   (Energy forms; power; units; conversions;
   Laws of Thermodynamics)

B. Energy Consumption
   1. History                                                           1, 8, 19
      (Industrial Revolution; exponential
      growth; energy crisis)
   2. Present global energy use                                         2, 19–21
   3. Future energy needs                                               2, 19–21

C. Fossil Fuel Resources and Use      19, 20
(Formation of coal, oil, and natural gas;
extraction/purification methods; world reserves
and global demand; synfuels; environmental
advantages and disadvantages of sources)

D. Nuclear Energy      20
(Nuclear fission process; nuclear fuel;
electricity production; nuclear reactor types;
environmental advantages/disadvantages;
safety issues; radiation and human health;
radioactive wastes; nuclear fusion)

E. Hydroelectric Power      15, 20
(Dams; flood control; salmon; silting; other
impacts)

F. Energy Conservation      13, 18, 19
(Energy efficiency; CAFE standards; hybrid
electric vehicles; mass transit)

G. Renewable Energy      2, 20, 21
(Solar energy; solar electricity; hydrogen
fuel cells; biomass; wind energy; small-scale
hydroelectric; ocean waves and tidal energy;
geothermal; environmental
advantages/disadvantages)

## VI. Pollution (25–30%)      Chapters 1, 5–7, 13–19, 22, 24

A. Pollution Types
  1. Air pollution      13, 17–19
(Sources—primary and secondary; major
air pollutants; measurement units; acid
deposition—causes and effects; heat
islands and temperature inversions; smog;
indoor air pollution; remediation and
reduction strategies; Clean Air Act and
other relevant laws)

  2. Noise pollution      13
(Sources; effects; control measures)

  3. Water pollution      5, 7, 15, 16, 19
(Types; sources, causes, and effects;
cultural eutrophication; groundwater
pollution; maintaining water quality;
water purification; sewage
treatment/septic systems; Clean Water
Act and other relevant laws)

  4. Solid waste      22
(Types; disposal; reduction)

B. Impacts on the Environment and Human
   Health
     1. Hazards to human health                      14, 17
        (Environmental risk analysis; acute and
        chronic effects; dose-response
        relationships; air pollutants; smoking and
        other risks)
     2. Hazardous chemicals in the environment     14, 22
        (Types of hazardous waste;
        treatment/disposal of hazardous waste;
        cleanup of contaminated sites;
        biomagnification; relevant laws)

C. Economic Impacts                         1, 6, 7, 24
     (Cost-benefit analysis; externalities;
     marginal costs; sustainability)

## VII. Global Change (10–15%)        Chapters 3, 4, 6, 11–12, 16–18

   A. Stratospheric Ozone                 17
      (Formation of stratospheric ozone;
      ultraviolet radiation; causes of ozone
      depletion; effects of ozone depletion;
      strategies for reducing ozone depletion;
      relevant laws and treaties)

   B. Global Warming                   3, 4, 6, 11, 18
      (Greenhouse gases and the greenhouse
      effect; impacts and consequences of global
      warming; reducing climate change;
      relevant laws and treaties)

   C. Loss of Biodiversity              3, 4, 11, 12, 16
     1. Habitat loss; overuse; pollution;
       introduced species; endangered and
       extinct species
     2. Maintenance through conservation     3, 4, 11, 12, 16
     3. Relevant laws and treaties             11

# Part II

# *A Review of Topics and Sample Questions*

Part II is keyed to *Environment: The Science behind the Stories*, Sixth Edition, and provides a focused review of the major environmental science topics. It gives an overview of the important information found in each text chapter and also provides sample multiple-choice and free-response questions.

Each Part II chapter summary is organized according to *"You Must Know"* and should be reviewed and studied before you attempt the practice multiple-choice and free-response questions provided. *After* answering the practice questions, be sure to thoroughly review the answers and explanations located in Part III, so that you prepare yourself for the range of questions and depth of information you will encounter on the AP Environmental Science Examination. This thorough approach will support a solid understanding and a confident ability to *apply and connect* the major concepts in the AP Environmental Science course.

TOPIC 1

# Environmental Science and Sustainability

## *Chapter 1: Science and Sustainability: An Introduction to Environmental Science*

**YOU MUST KNOW**

- The terms *environment*, *environmental science*, *renewable natural resources*, and *nonrenewable natural resources*.
- The importance of ecosystem services, with specific examples.
- The connection/interplay between population growth and resource consumption.
- The meaning of ecological footprint and how to calculate/assess its impact.
- The scientific method and the essentials of experimental design and data analysis.
- Concepts of sustainability and sustainable development, illustrated with specific examples.

***The terms* environment, environmental science, renewable natural resources, *and* nonrenewable natural resources**

Environment - all living and nonliving things

Environmental science - how the natural world works, how our environment affects us and how we affect it.

- The **environment** is a complex entity of all living and nonliving things, including, in its broadest sense, the structures, urban centers and other living spaces created by humans in the built environment.
- **Environmental science** is the interdisciplinary study (including natural and social sciences) of how the natural world works, how our environment affects us, and how we affect our environment—all using scientific methods to understand and solve environmental problems.
- **Renewable natural resources** are those that are replenished over relatively short periods. Sunlight, wind, and wave energy, for example, are perpetually replenished and essentially inexhaustible. Timber, water, and soil, however, can take months or decades to renew.

Renewable resources - resources that can be replenished over a short period of time } sun, wind, wave energy

*Nonrenewable
resources - resources
that cannot be
easily replenished
in short periods of
time (ex crude oil, natural gas)*

▮ **Nonrenewable natural resources** such as mineral ores and crude oil are finite in supply—once depleted, they are no longer available.

▮ Renewability of a natural resource is best viewed as a continuum, as even potentially renewable resources can be depleted if used faster than naturally replenished.

### *The importance of ecosystem services, with specific examples*

*Ecosystem services-
· purification of
air and water
· cycling of nutrients
· plant pollination
by animals
· waste recycling
systems*

▮ **Ecosystem services** are essential processes carried out by naturally functioning ecosystems, supporting living things and making human economic activity possible.

▮ Specific examples: the purification of air and water and essential cycling of nutrients by forests, plant pollination by animals, waste recycling systems provided by soil microbes, and global climate regulation by intact plant ecosystems.

### *The connection/interplay between population growth and resource consumption*

▮ Human population is currently over 7 billion, with 9 billion forecast by 2050.

▮ Agricultural revolution (~10,000 years ago) and industrial revolution (beginning in mid-1700s) have fueled human population growth.

▮ Population growth amplifies humans' resource consumption.

▮ Resource consumption exerts social and environmental pressures. By mining energy resources and manufacturing more goods, we are consuming more of the planet's limited resources. Although some of the world's population may increase their affluence, this is not universally true, often creating economic disparity.

### *The meaning of ecological footprint and how to calculate/assess its impact*

▮ An **ecological footprint** represents the total area of biologically productive land and water needed to produce the resources and absorb the wastes (e.g., $CO_2$ emissions) of a given person or population.

*Ecological footprint
↓
the total area of
biologically productive
land & water needed
to produce the
resources and absorb
the wastes of a
given person or
population*

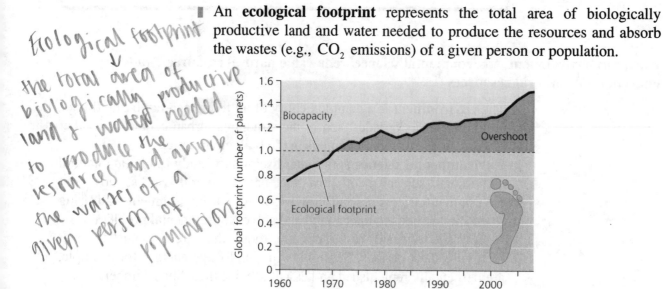

The Global Ecological Footprint and Its 50% Overshoot

▮ An **ecological footprint** can be calculated by answering specific questions regarding lifestyle and consumerism (www.footprintnetwork.org).

▮ Currently, humans are using more than 2.7 ha/person, creating a global ecological deficit of about 50%. That is, humanity is using renewable resources 50% faster than they are being replenished.

▮ Comparing human demand against biocapacity, it currently takes the natural resources of almost 1.5 planets to support humanity—demand is greatly exceeding nature's ability to provide ecosystem services.

▮ **Natural capital** is Earth's vast store of resources combined with the ecosystem services provided by the planet. This "capital" should be thought of as a bank account in which one leaves the principal intact and uses only the interest, in order to live sustainably. Currently we are drawing down Earth's natural capital faster than it is being renewed/replaced.

▮ The current U.S. ecological footprint is 7.2 ha/person (~18 acres) compared to China's 2.1 ha/person (5.2 acres) and India's 0.9 ha/person (2.2 acres).

*[handwritten margin note: Natural Capital – the Earth's vast store of resources combined with the ecosystem services provided by the planet.]*

*[handwritten margin note: World average Eco footprint – 2.7 ha. Largest is U.S.A ↓ 7.2 ha]*

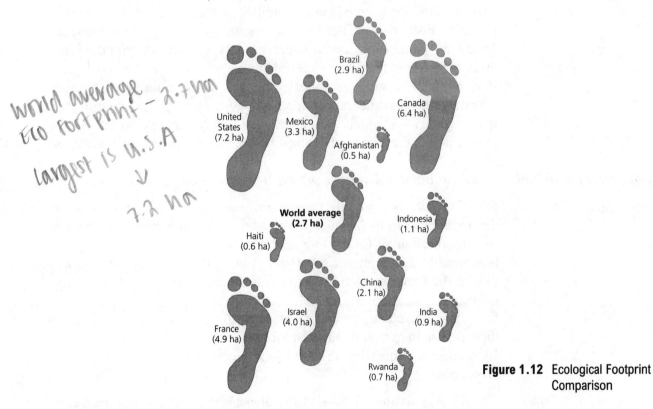

**Figure 1.12** Ecological Footprint Comparison

*The scientific method and the essentials of experimental design and data analysis*

▮ The **scientific method** is a technique for testing ideas by 1) making observations and 2) asking questions, which lead to 3) developing a **hypothesis** (a *testable* *explanation*).

▮ Using one's hypothesis to 4) generate a prediction that can be directly and unequivocally tested forms the basis of 5) scientific experimentation.

▊ A **controlled scientific experiment** is one that tests the validity of a prediction or hypothesis, manipulating one variable at a time—the **independent variable**. The **dependent variable** (that which the scientist measures) produces the data necessary to support or refute the hypothesis.

▊ The **control** of an experiment is often the unmanipulated case, used for comparison.

▊ **Replication**, that is, repeating an entire experiment and offering multiple replicates of the result of that experiment, is essential for adequate and accurate data generation.

▊ Scientists record information from their experiments, specifically valuing **quantitative data** (information expressed by numbers).

▊ An essential part of **data analysis** often involves choosing an appropriate graph format (line, bar, scatter plot, or pie chart) to visualize data.

▊ **Statistical tests** are always necessary to objectively and precisely determine the reliability of data.

▊ After multiple experiments are performed, results are analyzed and the findings are documented and submitted for publication in scientific journals. **Peer review** determines whether work merits publication. Following publication, other scientists may attempt to reproduce the original results in their own experiments.

▊ If a hypothesis survives repeated testing and continues to predict experimental outcomes accurately, it may potentially be incorporated into *a theory*—a widely accepted, well-tested explanation of one or more cause-and-effect relationships that has been extensively validated.

### Concepts of sustainability and sustainable development, illustrated with specific examples

▊ **Sustainability** is a guiding principle of modern environmental science that requires us to live in such a way as to maintain Earth's systems and its natural resources for the foreseeable future.

▊ **Sustainable development** satisfies our current needs without compromising the future availability of natural resources or our future quality of life.

▊ Currently, our consumption of natural resources has risen even faster than the rise in human population growth.

▊ Expanding and intensifying world food production has come at considerable cost.

1) Approximately 50% of the planet's land surface has been converted for agriculture.
2) Extensive use of pesticides and fertilizers has poisoned species, altered natural systems, and reduced Earth's biodiversity.
3) Erosion and poorly managed irrigation have destroyed much cropland.
4) Pollution from farmland/feedlots has polluted land, air, and water.

▌ Energy choices (especially regarding fossil fuel consumption) will greatly influence Earth's future.

▌ Sustainable solutions are increasing: Innovative renewable energy resources are being developed, energy-efficiency efforts are gaining ground around the world, agricultural reform and new environmental laws are reducing pollution and protecting the environment.

▌ The meaning of **sustainable development** can vary according to who is doing the developing; however, most people are now aiming to find solutions that foster economic advancement while also protecting the environment and promoting social equity—a concept known as the **triple bottom line**.

# *Chapter 24: Sustainable Solutions*

---

**YOU MUST KNOW**

- Specific approaches and examples of promoting sustainability.
- How to fully explain the concept of *modern* sustainable development.
- The connection between environmental protection and economic well-being.
- Strategies for sustainability—major approaches to building a sustainable future.

---

### *Specific approaches and examples of promoting sustainability*

*[handwritten: Ways to promote sustainibility]*

*[handwritten: Green buildings- constructed from sustainable building materials, built to reduce pollution]*

▌ **Audits** produce baseline data on how much an institution consumes and pollutes.

▌ **Recycling** and **waste reduction** are common sustainability efforts.

▌ **Green buildings** are constructed from sustainable building materials, are designed to reduce pollution, use renewable energy, and encourage efficiency in energy and water use. Buildings are responsible for 70–90% of a campus's greenhouse gas emissions, so building green is a current emphasis. *Leadership in Energy and Environmental Design* (LEED) standards are the agreed-upon standards guiding the design and certification of green buildings.

▌ **Water conservation**, especially important in arid regions, can include waterless urinals, "living machines" to treat wastewater, rain gardens for stormwater management, and low-flow showerheads.

▌ **Improving energy efficiency** can involve such measures as turning down hot water heater temperatures by 5 degrees, not powering unused buildings, and installing motion detectors to turn off lights when people are not in rooms.

▐ **Promoting renewable energy** can involve such measures as switching from fuel oil to carbon-neutral wood chips (reduces emissions by 40%), increasing solar power mix by installing PV solar system arrays, installing wind turbines, and purchasing "green tags" or carbon offsets (which subsidize renewable energy sources). Haverford College (PA) has achieved a 100% renewable energy status.

▐ **Reducing food waste** is being accomplished by campus dining services composting food scraps, employing trayless dining, and establishing "farm-to-campus" programs that provide locally grown produce.

▐ **Promoting more sustainable food consumption** involves buying more locally grown food and organic produce, and purchasing food in bulk with less packaging.

▐ **Promoting sustainable purchasing decisions** involves encouraging the purchase of energy-efficient appliances, recycled paper, certified sustainable wood, and nontoxic cleaning products; and eliminating herbicides and fertilizers used on lawns, gardens, and golf courses.

▐ **Sustainable transportation alternatives** include promoting alternative fuels (e.g., natural gas and biodiesel) and vehicles, bicycling, walking, and using public transportation.

▐ **Restoration of native plants, habitats, and landscapes** removes invasive species, enhances soil and water quality, and conserves water as vegetation is better adapted.

▐ **Carbon-neutrality as a major goal** involves reducing greenhouse gas emissions and/or employing carbon offsets to reduce the net amount of carbon released into the atmosphere to zero. College of the Atlantic (ME) was the first carbon-neutral campus.

▐ **Divestment from fossil fuel corporations** is a recent movement on many university campuses in which students are urging their administrations to divest from stock holdings of coal, oil, and gas industries. Unity College and Hampshire College became the first to formally divest in 2012.

### *How to fully explain the concept of* modern *sustainable development*

▐ **Sustainability** generally means sustaining human institutions and ecological systems in a healthy and functional state.

▐ Contributions of biodiversity and ecosystem goods and services are infinitely fundamental and valuable.

▐ Protecting the environment against the ravages of human development is no longer the sole consideration of sustainable development. Finding ways to promote social justice, economic well-being, and environmental quality, all at the same time (triple bottom line), is the goal of modern sustainable development.

### *The connection between environmental protection and economic well-being*

▌ There is a common misconception that economic well-being and environmental protection are in conflict.

▌ Protecting environmental quality can improve the triple bottom line, especially when we look beyond conventional economic accounting (which measures only private gain and loss) and include external costs and benefits that affect people at large.

▌ **Millennium Ecosystem Assessment** (2005) evidence shows how value is maximized by conserving natural resources rather than exploiting them for short-term profit.

  1)  value of intact wetlands > value of intensive farming
  2)  value of sustainable forestry > value of small-scale farming
  3)  value of intact mangroves > value of shrimp farming

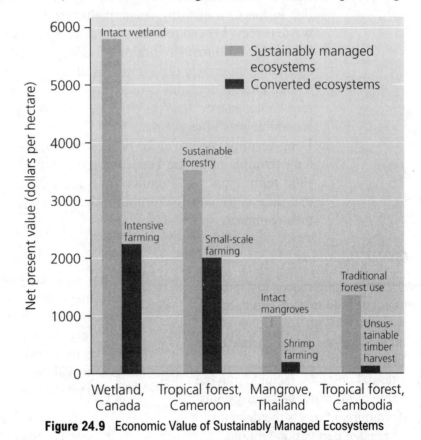

**Figure 24.9**  Economic Value of Sustainably Managed Ecosystems

▌ Environmental protection and green technologies can create rich sources of new jobs—especially in the renewable energy sector.

### *Strategies for sustainability—major approaches to building a sustainable future*

▌ **Consumer power:** Major approaches include expressing preferences through the political system, "voting with your wallet" for sustainably produced goods, and rethinking assumptions about economic growth as an ultimate goal.

■ **Population stability:** The growth in population and per capita consumption need to be halted; both are unsustainable and will eventually encounter serious limiting factors.

■ **Green technology:** Technology has traditionally increased environmental impact (I = PAT). However, new green technologies can help reduce impact: catalytic converters; recycling and wastewater treatment technologies; and solar, wind, and geothermal energy technologies.

■ **Local and global:** Promotion of local self-sufficiency and long-term thinking are critical criteria for sustainability.

■ **Political engagement**: Ordinary citizens have power as well, if they choose to exercise it. You can exercise your power at the ballot box; by attending public hearings; by donating to advocacy groups; and by writing letters, sending e-mails, and making phone calls to officeholders.

■ **Mimic natural systems** by promoting closed-loop industrial processes. For instance, transform linear pathways into circular ones, in which waste is recycled and reused, ultimately generating no waste.

■ **Systemic solutions:** Promotion of solutions that focus on "big picture" solutions, not merely the symptoms of the problem. A systemic solution to our energy demands would involve developing an array of clean renewable energy sources instead of looking for more immediate/accessible fossil fuel sources.

■ **Long-term perspective:** To be sustainable, a solution must work in the long term, as often the best long-term solution is not always the best short-term "political" solution. We must continue to build awareness that time is limited for turning around our increasing human environmental impact.

## Multiple-Choice Questions

1. *Modern* understandings of sustainable development hold that sustainability occurs when
   - (A) taking into account the triple bottom line goals of recycling.
   - (B) external costs are factored into the value of all converted ecosystems.
   - (C) development meets the demands of present generation without compromising future generations.
   - (D) environmental goals overlap with social and economic goals, achieving a triple bottom line.
   - (E) all economic costs of altering, developing, and maintaining converted ecosystems are taken into consideration.

2. In terms of ecosystem services, the sustainably managed ecosystems would
   - (A) provide enhanced and restored services over unaltered natural ecosystems.
   - (B) provide somewhat diminished or damaged services.
   - (C) provide more and more valuable services than converted ecosystems.
   - (D) have services that are different from but of equal value to those of converted ecosystems.
   - (E) restore the original biodiversity of ecosystems.

3. The world's average ecological footprint is 2.7 hectares/person compared to the U.S. average of 7.2 ha/person. The world is depleting renewable resources approximately ___% faster than they can be replenished; however, the U.S. impact alone is approximately ___ times greater than world average.
   (A) 50%; 3
   (B) 15%; 2.7
   (C) 30%; 5
   (D) 50%; 2.7
   (E) 5%; 30

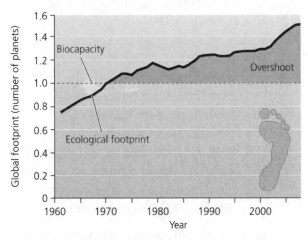

The Global Ecological Footprint and Its 50% Overshoot

4. How much larger was the global ecological footprint in 2005 than it was in 1985?
   (A) 10%
   (B) 15%
   (C) 30%
   (D) 50%
   (E) 1.1%

5. A controlled scientific experiment
   (A) often involves manipulating as many variables as possible in order to determine the factual basis of the relationship studied.
   (B) preferably involves collection of quantitative data over qualitative data.
   (C) is always designed to generate new scientific hypotheses and advance science.
   (D) is an activity designed to test the validity of a hypothesis.
   (E) involves manipulating only the dependent variable and measuring the independent variable.

6. Ecosystem services
   (A) contribute to keeping ecosystems economically productive.
   (B) are required to rebalance natural systems that we have disturbed.
   (C) are focused on providing humans with the essential life services.
   (D) are separate from economic value and exist along a continuum from renewable to nonrenewable resources.
   (E) are economically valuable services provided by natural systems.

7. A well-designed and well-controlled experiment includes
   (A) a testable hypothesis, one independent variable, replicates, and a quantifiable dependent variable.
   (B) a testable hypothesis, several independent variables, one replicate, and a quantifiable dependent variable.
   (C) a testable hypothesis, one independent variable, replicates, and several quantifiable dependent variables.
   (D) a testable hypothesis, a quantifiable independent variable, one replicate, and a qualitative dependent variable.
   (E) a testable hypothesis, one independent variable, a control, and a dependent variable that the scientist manipulates.

8. Which of the following will lead to a more sustainable world culture?
   (A) increasing green energy production for everyone
   (B) increasing consumer growth for all developing nations
   (C) creating industrial systems that are linear in design and operation
   (D) assisting developing countries to achieve the U.S. ecological footprint
   (E) reducing overall consumption and halting population growth

9. An ecological footprint is best described as showing
   (A) the major components of an average ecologist's footprint.
   (B) the impact of climate change on the planet.
   (C) the amount of carbon emitted on average by industrialized countries.
   (D) the total biologically productive land and water area required to sustain a person or population.
   (E) the total land area contaminated by the human population.

10. Structures given a LEED platinum ranking are
    (A) inefficient and need to be renovated and upgraded to meet green standards.
    (B) especially high in heavy metal content.
    (C) exemplary in sustainable design, especially energy efficiency.
    (D) constructed to function sustainably without electricity from fossil fuels.
    (E) free of asbestos, lead, and radon and therefore made of "green" construction materials.

11. Environmental science is most accurately viewed as a science
    (A) incorporating economics, social sciences, geography, and policy.
    (B) incorporating all the sciences, politics, economics, and social sciences.
    (C) incorporating world politics, decision making, and the environment.
    (D) designed to solve environmental problems using current technology.
    (E) designed to promote sustainable development on a global scale.

12. In a scientific experiment, the variable(s) which is/are manipulated is/are
    (A) the control, as this is the only quantifiable manner in which an experiment can be tested.
    (B) the independent variable.
    (C) the dependent variable.
    (D) both the independent and dependent variable, one at a time.
    (E) the dependent variable, in order to produce quantifiable results.

13. It has been well documented that world population growth negatively impacts ecosystem services. The current world population is estimated to be closest to
    (A) 3 billion.
    (B) 5.5 billion.
    (C) 7.4 billion.
    (D) 10 billion.
    (E) 6.9 million.

14. Specific examples of potentially renewable natural resources are
    (A) water, soil, air, and oil.
    (B) timber, water, and mineral resources.
    (C) agricultural crops, soils, and natural gas.
    (D) tidal energy, water, and mineral ores.
    (E) forests, water, and soil.

15. An example of a systemic approach to sustainability problem-solving would be to
    (A) avoid strip-mining in a national park or national forest.
    (B) use minimal herbicides and pesticides in industrial agriculture.
    (C) tax all industries contributing to acid precipitation falling throughout the northeastern United States.
    (D) research and employ diverse agricultural approaches that rely less heavily on chemical pesticides.
    (E) avoid overfishing in privately owned lakes because of biodiversity impacts.

16. Carbon-neutrality approaches refer to
    (A) reducing greenhouse gas emissions from fossil fuel combustion.
    (B) eliminating use of all carbon-based compounds.
    (C) cap-and-trade efforts currently employed by global enterprises.
    (D) adhering to the Montreal Protocol, purchasing carbon offsets, and employing ecosystem services.
    (E) becoming neutral with regard to renewable energy sources and negative with regard to nonrenewable energy sources.

17. Several of the least expensive ways to achieve a sustainable lifestyle would include
    (A) employing an energy audit, carbon-neutrality, and personal divestment from fossil fuel corporations.
    (B) recycling, water conservation, and purchasing carbon offsets.
    (C) purchasing compact fluorescent bulbs, using biodiesel transportation, and eating local produce.
    (D) employing water conservation, recycling, bicycling, and eating local produce.
    (E) using only renewable energy sources, recycling, and eating local produce.

*Free-Response Question*

> In 1997, a team of scientists led by University of Vermont ecological economist Robert Costanza published a landmark paper, *The Value of the World's Ecosystem Services and Natural Capital*. Their study determined that the monetary value of Earth's ecosystem services (from 16 major ecosystems including forests, grasslands, and all other terrestrial and aquatic ecosystems) totaled at least $33 trillion per year. Prior to this important study, forests and ecosystems were thought to have value primarily for their *economic services* (i.e., lumber, paper, and other wood products from harvesting forests). As an example of their intrinsic value, the Costanza study estimated that the world's forests alone are providing us with *ecosystem services* worth at least $4.7 trillion per year.

a) Other than the recreation value provided by forests, identify and explain two of the primary ecosystem services provided by global forests.

b) Assuming that the ecological valuation of forests is estimated to be 400% greater than that estimated by the economic gain of directly harvesting the timber, what is the approximate economic valuation per year?

c) In 1997, the year this report was published, the global gross national product (the monetary value of all good and services produced world-wide) was $18 trillion. In view of this fact, discuss with specifics the meaning of *sustainability* and its implications in the context of the Constanza study.

# Principles of Ecology

## Chapter 2: Earth's Physical Systems: Matter, Energy, and Geology

---

**YOU MUST KNOW**

- The fundamentals of matter and chemistry and be able to explain their application to real-world situations.
- How water's unique chemistry supports the living world.
- The role of organic compounds in ecosystem function.
- How to differentiate among the basic forms of energy and explain the first and second laws of thermodynamics.
- The basics of photosynthesis, cellular respiration, and their importance to living things.
- How the basic geology processes of plate tectonics and the rock cycle shape the earth's landscape.
- Major types of geologic hazards and the approaches required to mitigate their impacts.

---

*The fundamentals of matter and chemistry and their application to real-world situations*

▌ **Environmental science**, the broadest of all scientific disciplines, must include the study of **matter** (that which takes up space and has mass) in order to understand the underlying nature of environmental problems.

▌ **Chemistry** (the study of elements and compounds and how they interact) plays a central role in understanding most environmental issues. Examples:

  ▬ How do gases such as $CO_2$ and $CH_4$ contribute to global climate change?

  ▬ How do pollutants such as $SO_2$ and NO cause acid rain?

  ▬ How do pesticides affect the health of wildlife and humans?

■ **Law of conservation of matter:** Matter cannot be created or destroyed, but can be transformed into new substances.

■ The amount of matter on Earth is constant and is recycled in biogeochemical cycles.

■ Matter cannot be "thrown away" or made to disappear, even though it can change form.

■ **Elements** (composed of only one type of atom and cannot be broken down into substances with other properties) and **compounds** (composed of two or more elements chemically combined in a fixed ratio) make up all matter.

■ The elements C, H, O, and N make up 96% of living matter.

■ **Isotopes** are atoms of the same element with a differing number of neutrons and mass.

■ **Radioisotopes** change chemical identity as they shed subatomic particles, emitting radiation and decaying at a rate determined by their **half-life** (the amount of time it takes for one-half the atoms to give off radiation and decay). The 2011 Fukushima Daiichi nuclear plant accident released numerous types of radioisotopes (Cs-137 an example), contaminating the Pacific Ocean waters, sediment, and marine life for decades to come.

■ **Ions** are charged atoms; their charge is due to an unequal number of protons and electrons. The changing radiation emitted by radioisotopes is called **ionizing radiation** because of the ions generated when striking molecules, affecting the stability and functionality of biological molecules such as DNA.

### *How water's unique chemistry supports the living world*

■ The **structure of the water molecule** is the key to its special properties: The two hydrogen atoms and the one oxygen atom bond together to form a **polar molecule**—the end bearing the oxygen has a slightly negative charge, whereas the end bearing the hydrogen atoms has a slightly positive charge.

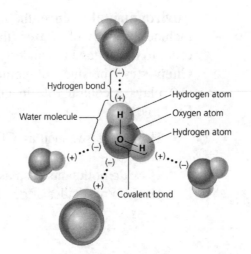

**Figure 2.4** The Water Molecule and Its Hydrogen Bonds

▪ Because of water's polar nature, **hydrogen bonds** form between water molecules, giving the properties of **cohesion** (water molecules "sticking" to one another), a **high ability to dissolve** many substances, and a **high specific heat** (water can absorb a great deal of heat with only small changes in its temperature). These are properties that help to support life and stabilize Earth's climate.

▪ Some water molecules split apart, forming hydrogen ions ($H^+$) and hydroxide ions ($OH^-$). Solutions in which the concentrations of $H^+ > OH^-$ are **acidic (pH < 7)**; whereas solutions in which $OH^- > H^+$ are **basic (pH > 7)**.

▪ Pure water is neutral (pH of 7) and $[H^+] = [OH^-]$.

▪ The pH scale is logarithmic; each step from 0 to 14 represents a tenfold difference in hydrogen ion concentration.

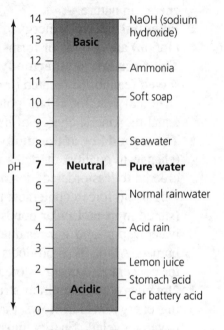

**Figure 2.5**  The pH Scale

▪ Industrial air pollution makes precipitation more acidic. The pH of rain throughout most of the eastern United States now averages well below 5. The oceans worldwide are also becoming more acidic as seawater absorbs excess $CO_2$ from fossil fuel emissions.

## *The role of organic compounds in ecosystem function*

▪ **Organic compounds**, always composed of carbon and hydrogen atoms, and sometimes the elements N, O, S, and P, are essential to all living things. Inorganic compounds lack carbon–carbon bonds.

▪ **Macromolecules** (organic compounds including proteins, nucleic acids, carbohydrates, and lipids) play key roles as energy sources (especially glucose: $C_6H_{12}O_6$) and are building blocks of living organisms.

- **Hydrocarbons** (organic compounds consisting solely of carbon and hydrogen atoms) are important to environmental science because they are the primary constituents of fossil fuels and all petroleum products.
- **Synthetic polymers (plastics)** are often long-lasting, resistant to chemical breakdown, and play a large role in our manufactured products. These same characteristics also result in problems for wildlife, human health, water quality, and waste management.

### How to differentiate among the basic forms of energy and explain the first and second laws of thermodynamics

- **Energy** is the capacity to accomplish work and change position, physical composition, or the temperature of matter. Geological events (earthquakes and tsunamis) involve some of the most dramatic releases of energy in nature.
- **Potential energy** (energy of position) and **kinetic energy** (energy of motion) are two major forms.
- **The first law of thermodynamics** states that energy is always conserved (cannot be created or destroyed) but can change in quality during chemical reactions. The potential energy of water behind a dam will equal the kinetic energy of its eventual movement downstream.
- **The second law of thermodynamics** focuses on the conversion of energy (change in quality). It states that energy will always change from more-ordered to less-ordered, degrading in quality and increasing in entropy, so long as no force counteracts this tendency. The potential energy of wood (stored in its molecular bonds) transforms into heat, light, carbon dioxide, water vapor, and the residue of carbon, ash, and particulates after it is burned. All of these products are less-ordered and higher in entropy than the concentrated energy stored in the original wood.
- Organisms maintain their structure and function by continually consuming energy, which flows through trophic levels in food chains and is never recycled. Energy must be constantly supplied to an ecosystem, usually by the sun.
- **Energy conversion efficiency** is the ratio of useful output of energy to the amount needed for input.
- **Energy transfer/flow** between trophic levels is **typically** only **10% efficient**.

### The basics of photosynthesis, cellular respiration, and their importance to living things

- **Photosynthesis** is the process by which **autotrophs/primary producers** make organic compounds by converting light energy from the sun into chemical energy (sugars/food). Sunlight, life's primary energy source, powers the autotroph's chemical reactions, converting $CO_2$ and $H_2O$ into the concentrated chemical energy of sugars ($C_6H_{12}O_6$), their food.
- **Cellular respiration** is the process by which most organisms release the chemical energy of sugars ($C_6H_{12}O_6$) to provide energy for both

plants' and animals' cellular activities. $CO_2$ and $H_2O$ are also products of respiration. As in all energy conversions, the energy released per glucose molecule is only about two-thirds of energy input per glucose molecule; therefore, the energy conversion is not 100% efficient.

▌ Organisms at each trophic level use most of the food energy for their own life's needs (building new tissue, movement, essential chemical reactions, etc.).

▌ Only about 10% of the original energy at any trophic level is available to be passed up to the next higher trophic level, creating a pyramid of energy that is never inverted.

▌ **Geothermal energy** is energy generated deep within Earth as a result of radioactive decay, producing pressure and heat. Where this energy heats groundwater, natural surface eruptions of steam can result.

### *How the basic geological processes of plate tectonics and the rock cycle shape the earth's landscape*

▌ Earth's geology is dynamic, with very long periods of time involved in geologic processes, especially when compared to the length of a human lifetime.

▌ Earth is composed of distinct layers (**crust, mantle,** and **core**) that differ in composition, temperature, and density.

▌ **Plate tectonics**, the slow/large-scale movement of Earth's lithospheric plates (uppermost mantle), is a fundamental process that shapes Earth's physical geography, also producing earthquakes and volcanoes along the various types of plate boundaries (divergent, transform, and convergent).

▌ Tectonic movements build mountains, islands, and continents, and shape ocean geography; therefore, they are highly influential in **biogeography** (geographic distribution of life).

▌ Over geologic time, rocks, and the minerals that comprise them, are heated, melted, cooled, and reassembled in a very slow process called the **rock cycle**, which recycles all matter within the lithosphere.

### *Major types of geologic hazards and the approaches required to mitigate their impacts*

▌ By driving plate tectonics, Earth's geothermal heating gives rise to the creative forces shaping our planet, some of which pose hazards to humans.

▪ Earthquakes result from movement at transform and convergent plate boundaries that occur around the Pacific "ring of fire." The 2011 Tohoku earthquake resulted from movement along two converging plates (the Pacific and Eurasian plates).

▪ Increased extraction of crude oil and natural gas from conventional wells and hydraulic fracturing has resulted in the increased injection of wastewater into porous rock layers wherever this extraction has occurred.

- Since 2009, the south central United States, particularly Oklahoma, has experienced rapid increase in earthquake activity near the injection sites.
- Scientific studies in 2013 have now confirmed the link between wastewater injection and increased seismic activity in this region.
- Although governments now monitor many wastewater injection wells for pressure, and mandate slower injection rates and volumes, studies indicate a possible two-decade-long lasting effect for increased seismic activity.
- **Volcanoes** arise from rifts (divergent plate boundaries), subduction zones (convergent plate boundaries), or hotspots: the Hawaiian Islands provide an example of a series of islands resulting from a tectonic plate moving across a volcanic hotspot.
- Tsunamis (caused by earthquakes displacing large amounts of ocean water) can cause immense human and environmental damage when this powerful surge of seawater displaces large volumes of ocean-bottom sediment, pushing water upwards into waves that spread outward from the earthquake site.

- Humans can reduce negative impacts/mitigate natural hazards through better land use practices by understanding Earth's geological processes.

# Chapter 3: Evolution, Biodiversity, and Population Ecology

---

**YOU MUST KNOW**

- How evolution by natural selection generates and shapes biodiversity.
- Reasons for current species extinction and Earth's known past mass extinction events.
- Characteristics that make species vulnerable to extinction: K-strategists versus r-strategists; specialists versus generalists.
- Levels of ecological organization, including habitat and niche.
- Population characteristics used to predict population growth potential, growth rates, and survivorship curves.
- Differences between exponential and logistic growth models.
- Differences between density-dependent and density-independent factors.
- Specific challenges and current efforts involved in conserving biodiversity.

*How evolution by natural selection generates and shapes biodiversity*

▌ Darwin's **theory of natural selection**: Organisms often produce excess young; individuals within each species vary in their traits; some individuals will be better suited to their environment than others and the genes/traits of these individuals that enhance survival will be passed on and become more prominent in future generations.

▌ **Mutations** and **sexual reproduction** produce the essential genetic variation upon which the process of natural selection acts.

▌ **Selective pressures** from the environment influence adaptation. Environmental conditions determine what pressures natural selection will exert, and these selective pressures affect which members of a population will survive and reproduce.

▌ **Artificial selection** is the process of trait selection conducted under human direction, rather than environmental pressures, and has given rise to the wide diversity of food crops. Corn, wheat, rice, and most of the entire agricultural system are based on artificial selection.

▌ **Speciation** is the process by which new species are generated. Biological diversity, or **biodiversity**, refers to the variety of life across all levels of biological organization, including the diversity of species, genes, populations, and communities.

▌ **Allopatric speciation**, thought to be the main mode of speciation, occurs when an original population separates into *geographically isolated* populations that do not interbreed. As each separate population experiences differential selective/environmental pressures over time, it is possible that significant genetic differences will accumulate between these populations because of their differing environments. Allopatric speciation is confirmed when individuals from these formerly separated populations come back together and are unable to mate and produce viable offspring. The 18 living species of Hawaiian honeycreeper are examples of allopatric speciation. As new volcanic islands emerged and as forests expanded and contracted over the millennia, the original ancestral species population was split many times and came to live in new separated environments. The various honeycreeper species diverged from one another in this manner, evolving different body shapes, color, sizes, and unique feeding behaviors based on bill shape/length.

▌ **Sympatric speciation** occurs when a small part of a population becomes a new species because of reproductive isolation within the *same* geographic area. For example, populations of some insects may become reproductively isolated if they feed and mate exclusively on different types of plants in the same geographic area.

▌ ~ 3.5 billion years of life on Earth have culminated in an estimated 3 million to 100 million species living today, with only 1.8 million species currently discovered and described. The vast majority of all species have become extinct over time.

▌ Speciation and extinction together determine Earth's current biodiversity, with **phylogenetic trees** providing an illustration of the relationships of these species and the history of life's divergencies and evolution.

### Reasons for current species extinction and Earth's known past mass extinction events

▌ Species often become extinct when environmental conditions change rapidly or severely enough that a species cannot genetically adapt to the change—the slow process of natural selection often does not have sufficient time to work.

▌ **Small populations** often lack the genetic variation required to protect them against environmental change.

▌ **Island-dwelling species** and other **small range populations** are particularly vulnerable because of the small areas and isolation by water, not allowing species movement if severe changes occur in their local environment. Half of Hawaii's native birds became extinct soon after humans introduced non-native rats, cats, and mongooses.

▌ **Mainland "islands" of habitat** (such as remnant prairie surrounded by agricultural lands) can host endemic species that are vulnerable to extinction.

▌ There have been five **mass extinctions**, the best-known occurring 66 million years ago (the K-T event), bringing an end to the dinosaurs. Evidence suggests that the impact of a gigantic asteroid caused atmospheric soot to blot out the sun, inhibiting photosynthesis and plant growth, causing plants to die, and animals to starve. These mass extinction events are thought to have wiped out 50–95% of our planet's species each time.

▌ During Earth's 4.6-billion-year history, there has been climate change, the subsequent rise and fall of sea level, the evolution of new species (adapting to gradually changing environments), followed by subsequent interspecific competition among species.

▌ Human impact (especially the alteration of and destruction of natural habitats due to population growth, industrial development, and resource extraction and use) has profoundly affected current extinction rates, which the Millennium Ecosystem Assessment Report estimates to be 100–1,000 times higher than background extinction rates. Many biologists have concluded that Earth is currently entering its **sixth mass extinction** as a result of these diverse and negative human impacts.

### Characteristics that make species vulnerable to extinction: K-strategists versus r-strategists; specialists versus generalists

▌ **K-strategists** are species having few offspring that are larger in size, require longer parental care, and reach maturity at a later age.

▌ **r-strategists** are species devoting energy and resources to producing many offspring in a relatively short time, with little or no parental care, that are small in size and come to maturity early.

▌ **Specialists** are species that can survive only in a narrow range of habitats that contain very specific resources. Because of these limitations on their specific way of life, they are more vulnerable to extinction when conditions change and threaten the habitat or resource on which they have specialized. The native Hawaiian forest birds, the various honeycreeper species, are a good example because each species has a unique bill shape exquisitely adapted for feeding on various insects/larvae that live on/in the wood of the native Hawaiian trees.

▌ **Endemic species** are those that occur nowhere else on the planet but in a very specific region. These species also face increased risk of extinction because their members often belong to a single, sometimes small, population. The Hawaiian state bird, the nene, is an example.

▌ **Generalists** are species with broad tolerances, able to use a wide array of resources, and succeed by being able to live in many different places with variable conditions. These species are often able to endure environmental change better than specialists, making them less susceptible to extinction; however, they may not thrive in any one situation as much as a specialist would. The common myna, a bird introduced to Hawaii from Asia, is an example of a generalist. Its unremarkable bill allows it to eat many types of food in many types of habitats, facilitating its spread through all areas where human development has altered the Hawaiian landscape.

### *Levels of ecological organization, including habitat and niche*

▌ **Ecology:** the scientific study of interactions among organisms and between organisms and their environments, hierarchically organized from individual **organisms**, to **populations** (a group of individuals of same species that live in a particular area), to **communities** (a set of populations of different species living together in a particular area), to **ecosystems** (a community plus its nonliving environment and all interactions between them), and ultimately to the **biosphere** (the sum total of living things on Earth and all areas inhabited by life).

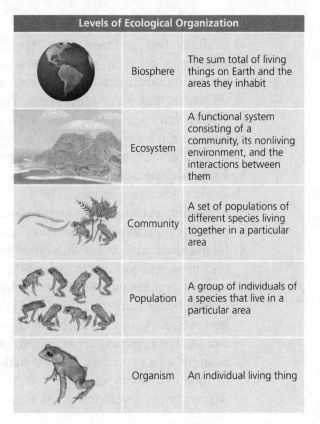

| Levels of Ecological Organization | | |
| --- | --- | --- |
| | Biosphere | The sum total of living things on Earth and the areas they inhabit |
| | Ecosystem | A functional system consisting of a community, its nonliving environment, and the interactions between them |
| | Community | A set of populations of different species living together in a particular area |
| | Population | A group of individuals of a species that live in a particular area |
| | Organism | An individual living thing |

Hierarchical Levels of Life

▌ **Habitat:** the living and nonliving elements surrounding where an organism lives—its "address."

▌ **Niche:** an organism's use of resources and its functional role in a community, including habitat use, food consumption, role in energy flow/matter cycling, and all interactions with other organisms—its "profession."

▌ **Ecology** is studied at several levels: population, community, and ecosystem.

1) **Population ecology** examines the dynamics of population change and helps us understand why populations of some species (such as endangered honeycreepers) decline, while populations such as humans increase.

2) **Community ecology** focuses on patterns of species diversity and on interactions among species.

3) **Ecosystem ecology** reveals patterns such as energy flow and nutrient cycling by studying both living and non-living components of the ecosystem. Today's studies of how climate change is affecting various ecosystems around the world are examples of this level.

***Population characteristics used to predict population growth potential, growth rates, and survivorship curves***

▌ **Population growth potential** attributes:

1) population size
2) density: number of individuals per unit area or volume
3) distribution: spatial arrangement (random, clumped, or even)
4) sex ratio: proportion of males to females
5) age structure: relative number of individuals of each age group
6) population: birth/death rates

▌ **Population growth rate:** to obtain an overall population growth rate, the total rate of change in a population's size per unit time, one must also take into consideration the effects of migration.

- ▪ Crude birth rates (CBR) and crude death rates (CDR) are expressed per 1,000 individuals per year.
- ▪ (CBR − CDR) + (immigration rate − emigration rate) = population growth rate
- ▪ (births + immigration) − (deaths + emigration) ÷ overall population size = growth rate

▌ **Survivorship curves:**

1) Type I shows low death rates during early and midlife, followed by death rates that increase sharply in older age groups.
2) Type II survivorship curves show a constant death rate over a life span.
3) Type III curves show very high death rates for young, then a lower death rate for the few individuals surviving to older age.

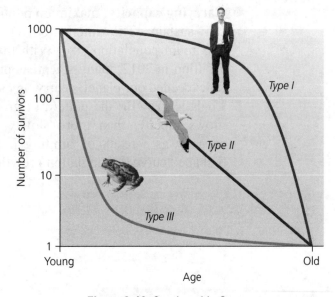

**Figure 3.19** Survivorship Curves

## *Differences between exponential and logistic growth models*

▌ **Exponential growth:** occurs when a population increases by a fixed percentage each year and is pictured as a J-curve—growth under ideal conditions. It occurs in nature when a population is small, competition is minimal, and the organism is introduced into a new environment with abundant resources. The Eurasian collared dove is currently spreading across North America, propelled by exponential growth.

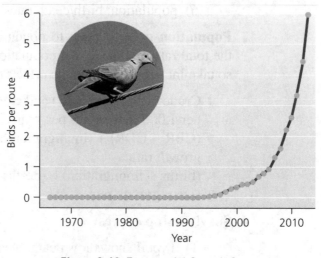

**Figure 3.16** Exponential Growth Curve

▌ **Limiting factors:** the physical, chemical, and biological attributes of the environment that restrain population growth, restricting indefinite exponential growth.

▌ **Carrying capacity:** maximum population size that a given environment can sustain. The interaction of limiting factors determines this maximum population size. With the global human population passing 7 billion in 2012, many scientists presented data showing that we have far exceeded our planet's carrying capacity for humans.

▌ **Logistic growth:** the pattern of growth in which a population initially grows rapidly, then more slowly, finally stabilizing at its carrying capacity as a result of limiting factors. This pattern is shown by an S-shaped curve of population growth over time.

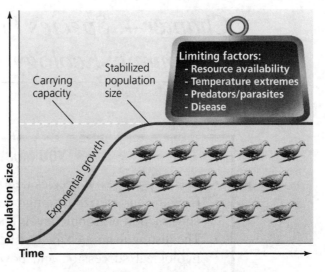

**Figure 3.17** Logistic Growth Curve

### *Differences between density-dependent and density-independent factors*

▊ **Density-dependent factors** are involved when death rates rise or birth rates fall as the overall population density rises. Examples include diseases and predation.

▊ **Density-independent factors** are involved when death rates do not change with increase or decrease in population density. Examples include temperature extremes, floods, and fires.

### *Specific challenges and current efforts involved in conserving biodiversity*

▊ **Introduced species** pose challenges for all native populations and communities. Island species are particularly vulnerable, as the Hawaiian islands have demonstrated (i.e., the negative impacts from cattle, goats, sheep, pigs, alien grasses, shrubs, and trees have eliminated half [70 of 140 species] of all native Hawaiian bird species).

▊ Pressures from human development, resource extraction, and human population growth are speeding the rate of environmental change and altering habitats across the planet.

▊ The creation of national parks, followed by ecotourism, help preserve biodiversity but cannot prevent all species extinction. Fundamental human behavior change with regard to energy and resource consumption needs to be taken more seriously to prevent further erosion of Earth's biodiversity.

▊ **Climate change** now poses an additional challenge, as even with preserving new tracts of land designated as protected areas, temperature increases and shifting rainfall patterns cause areas within these protected areas to become unsuitable for the species they were meant to protect. As global temperatures rise, Hawaii's bird populations are especially vulnerable, as mosquitoes and malaria are able to move upslope toward protected wildlife refuges.

# Chapter 4: Species Interactions and Community Ecology

**YOU MUST KNOW**

- The major types of species interactions and how they compare.
- How to characterize feeding relationships and energy flow in the construction of trophic levels and food webs.
- The distinguishing characteristics of keystone species.
- Ecological succession and how communities can change over time.
- How to predict the potential impacts of invasive species, including specific responses to biological invasions.
- Restoration ecology's goals and methods.
- The defining characteristics and significance of the major terrestrial biomes.

### *The major types of species interactions and how they compare*

▌ **Competition** occurs when multiple organisms seek the same limited resource.

  - ▪ **Intraspecific:** competition between members of the same species.
  - ▪ **Interspecific:** competition between two or more different species.
  - ▪ **Fundamental niche:** the full niche of a species.
  - ▪ **Realized niche:** that niche which is displayed when, as a result of competition, an individual plays only part of its role or uses only part of its resources.
  - ▪ **Resource partitioning:** the evolutionary process by which species adapt to competition by using slightly different resources and/or using them in slightly different ways.

▌ **Predation** is the process by which individuals of one species hunt, capture, kill, and consume individuals of another species—the prey. Predation is a primary organizing force of community ecology (feeding relationships) and sometimes drives **cyclical population dynamics** (increases and decreases in the population of one species driving increases and decreases of another, such as hare and lynx population cycles).

▌ **Parasitism** is the interaction where one species derives benefit by harming (but usually not killing) another species: parasitic sea lamprey and lake trout.

- **Mutualism** is the interaction where both species benefit, each providing some resource or service that the other needs: honeybees and flowering plants.
- **Commensalism** is the interaction where one species benefits and the other species is unaffected: remora and shark.
- **Herbivory** occurs when animals feed on plants; this is one of the most common species interactions in ecosystems.

### *How to characterize feeding relationships and energy flow in the construction of trophic levels and food webs*

- Energy is transferred in food chains between trophic levels according to the **10% law**, with each trophic level containing just one-tenth the energy of the trophic level below it.

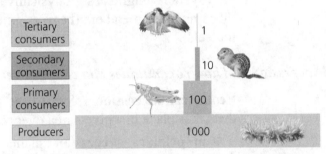

**Figure 4.10**  Pyramid Relationship of Energy and Biomass

- **1st trophic level** contains producers, or autotrophs (those species capable of photosynthesis).
- **2nd trophic level** contains **primary consumers**, or herbivores.
- **3rd trophic level** contains **secondary consumers**, or carnivores.
- **4th trophic level** contains **tertiary consumers**, or top carnivores.
- **Detritivores** (millipedes, soil insects) scavenge dead organisms, feces, dead leaves, and wood, converting organic matter to inorganic compounds.
- **Decomposers** (bacteria, fungi) break down non-living matter into simpler constituents, recycling nutrients back to the producers.

- **Biomass** refers to the organic material making up a living organism(s). When quantitatively determining biomass for ecological studies, it is the total dry weight of the organism(s).
- Lower trophic levels always contain the most energy and generally contain the most biomass and number of individuals.
- A pyramid-shaped diagram best describes the energy flow between trophic levels in a community. The pyramid pattern illustrates why eating at lower trophic levels (vegetarianism) decreases a person's ecological footprint—because when we eat animals, we use up far more energy per calorie than we gain by eating plants directly.

### The distinguishing characteristics of keystone species

- **Keystone species** have strong or wide-ranging impacts in their ecosystem, far out of proportion to their actual abundance.
- **Top predators** (wolves, sea stars, sharks, sea otters) are frequently considered keystone species and can promote **trophic cascades** (keeping species in intermediate trophic levels in check, while promoting species in lower trophic levels).
- Sea otter, sea urchin, kelp forest relationships in the Pacific Northwest were extensively studied in the 1990s. Current sea otter declines are hypothesized to be caused by *Orcas* (killer whales), which, deprived of their usual food sources (other whale species, harbor and fur seals), are now feeding on sea otters; this development has set off a trophic cascade (otters decrease, urchins increase, kelp decrease).
- **"Ecosystem engineers"** physically modify the environment shared by community members (beavers, prairie dogs, ants, zebra, and quagga mussels).

### Ecological succession and how communities can change over time

- **Ecological succession** refers to the transition in species composition that takes place in a given area over ecological time.
  - In **primary succession**, plants and animals gradually invade a region that is devoid of life and where soil has not yet formed, such as after a glacier retreats or after volcanic lava/ash spreads across a landscape. **Pioneer species** (lichens) begin this soil-building process, eventually allowing grasses, shrubs, and the trees associated with the region's climax community to increase.
  - In **secondary succession**, the existing community has been cleared or disturbed by fire, landslide, or farming; however, the soil is intact. An abandoned farm field will eventually return to a more diverse community of grasses, shrubs, and trees if no further disturbance occurs.

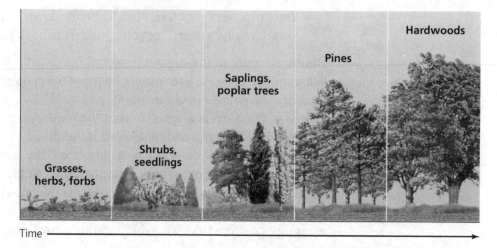

Secondary Succession Occurs After Environmental Disturbance

- Communities may undergo **phase** or **regime shifts**, involving irreversible change if environmental disturbance is severe enough. An example is the many coral reef communities that have experienced the loss of herbivorous keystone fish species (due to overfishing). These reefs are now overgrown with algae and may never shift back to the original coral-dominated state.

- Communities are now viewed most accurately as temporary/fluid associations of individual species, each species responding independently to their own limiting factors, the complexities of introduced species arriving into the community, and other complex factors.

- Ecological studies after the Mount St. Helens volcanic eruption in Washington (1980) are providing new insights into the complexities and unpredictable nature of succession. Biological communities appear to recover from disturbance in ways that are dynamic and highly dependent on chance, not the linear path from pioneer species to hardwood forests once predicted.

### *How to predict the potential impacts of invasive species, including specific responses to biological invasions*

- **Introduced species** are non-native species, introduced (accidentally or deliberately) into a new area, often spreading widely, rapidly becoming dominant, with the potential for interfering with a community's normal functioning.

- The zebra mussel demonstrates the variety of unexpected economic and environmental impacts that can result.

  - After zebra mussel introduction, phytoplankton and small zooplankton decline (zebra mussels are filter feeders) in aquatic ecosystems.
  - Larvae and juveniles of open-water fish then decline (they also depend on phytoplankton and zooplankton).
  - Littoral (shallow-water) fish tend to increase as zebra mussel shells provide habitat structure and their feces provide nutrients for many benthic species.

- Introduced species may become **invasive species** when limiting factors that regulate their growth are absent (no predators and/or fewer competitors) and their population numbers increase dramatically (Chestnut blight fungus, European grasses introduced to the American west, zebra and quagga mussels throughout most American waterways, and kudzu in the American south).

- Total eradication of invasive species (involving physical removal and/or applying toxic chemicals) is often so difficult that managers aim merely to control populations by limiting their spread or impact.

- By analyzing the biology of a species, scientists can model the conditions under which it will survive, thus predicting where it might spread.

## Restoration ecology's goals and methods

▌ **Restoration ecology** is the study of the historical conditions present in ecological communities as they existed before human alteration.

▌ Restoration ecology's specific goals/methods include reestablishing a wetland's ability to filter pollutants or recharge groundwater (Florida Everglades restoration) and reestablishing native prairie vegetation by weeding out invasive species and introducing controlled burning to mimic fires that historically maintained the community (tallgrass prairie restoration).

## The defining characteristics and significance of the major terrestrial biomes

▌ **Biomes** are the major regional complexes of similar biological communities, often recognized by the indicator plants and animals occupying each.

▌ **Climate**, especially precipitation and temperature, determines the plant life in each biome, which in turn determines the animal life. Major terrestrial biomes include the following.

■ **Temperate deciduous forest:** fertile soils, deciduous trees (oaks, maples), and precipitation spread evenly throughout the year.

■ **Deserts:** very sparse rainfall, extreme temperatures (hot or cold), and all plants and animals demonstrating water conservation adaptations.

■ **Temperate grassland (prairie):** marked by too little rainfall to support trees, but with fertile soils, occasional fires, and large grazing mammals.

■ **Tropical rainforest:** great biodiversity growing under constant, warm temperatures and high rainfall, but with nutrient-poor and thin soils.

■ **Tundra:** marked by permafrost, cold temperatures, and little rainfall, producing a landscape of lichens and low vegetations, with no trees.

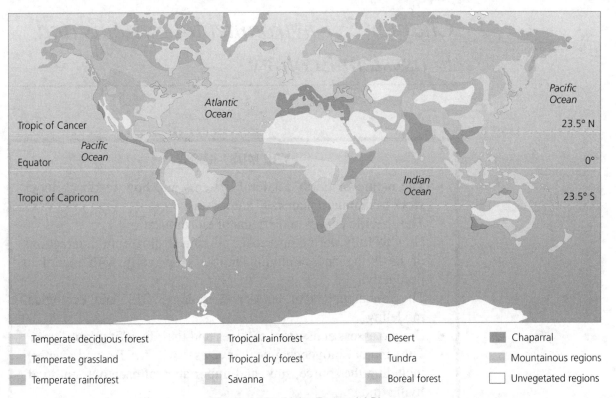

Temperate deciduous forest    Tropical rainforest    Desert    Chaparral

Temperate grassland    Tropical dry forest    Tundra    Mountainous regions

Temperate rainforest    Savanna    Boreal forest    Unvegetated regions

**Figure 4.16**    Major Terrestrial Biomes

- **Boreal/coniferous forest (taiga):** dominated by cone-bearing plants; acid/nutrient-poor soils; long, cold winters; and moderate precipitation.
- **Savanna:** tropical grassland marked by high temperatures and distinct rainy seasons, with grazing animals concentrated near water holes.

❚ Altitude increases can create changes of precipitation and temperature, analogous to latitude changes, producing biomes ranging from desert to tundra.

❚ Understanding the reasons why biomes occur where they do allows for ecological restoration, which strives to undo some of the negative effects of human impact.

# Chapter 5: Environmental Systems and Ecosystem Ecology

## YOU MUST KNOW

- The nature of Earth's natural systems and how feedback loops work to maintain planetary homeostasis.
- The specific causes and effects of eutrophication.
- How to fully define an ecosystem and evaluate the interactions of its living and nonliving entities, especially with regard to productivity.
- The fundamentals of landscape ecology, GIS, and ecological modeling.
- How to assess ecosystem services and their benefit to our lives.
- The major biogeochemical cycles (C, N, P, and water), including the source, sink, and importance of each nutrient to all living things.
- How human activities adversely affect biogeochemical cycles, and possible approaches for improved resource management.

### The nature of Earth's natural systems and how feedback loops work to maintain planetary homeostasis

- Earth's environment consists of complex networks of interlinked systems. These **systems** consist of both physical and biological relationships interacting with one another through the exchange of energy, matter, and information.
- When all these planetary systems are working together naturally (*without* the effects of human intervention), natural processes will move in opposing directions at equivalent rates so that their effects create a balance, a state known as **dynamic equilibrium**. This tendency of a system to maintain constant or stable conditions is referred to as **homeostasis**, and is a critical process to normal functioning of organisms, communities, ecosystems, and the biosphere.
- The Chesapeake Bay estuary and the Mississippi River watershed are examples of two complex systems with many interrelationships. Their interconnected environmental problems demonstrate the complexity involved in the analysis of and solutions to these problems.
- For centuries, the Chesapeake Bay oysters kept this estuary's waters clear by filtering nutrients and phytoplankton from the water column. However, by 2010, the oyster fishery was reduced to 1% by overhar-

vesting, habitat destruction, and agricultural pollution from excess nitrogen and phosphorous fertilizers. These complex and interrelated problems resulted in many hypoxic "dead zones" throughout this vast estuary. Now, strict pollution budgets aim to reduce N and P inputs, combined with oyster restoration efforts, in an attempt to restore the Chesapeake Bay to its former economic and biological productivity.

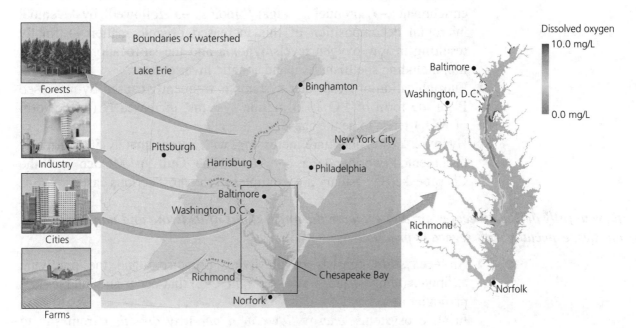

**Figure 5.3**  Chesapeake Bay Watershed

▌ **Hypoxia** (low dissolved oxygen) in the Chesapeake Bay's estuary, resulting from excessive N and P fertilizer use in the adjoining watershed, illustrates one result of these complex system interactions.

▌ Systems consist of interacting components that generally involve feedback loops. A **negative feedback loop** occurs when a system's output acts as an input that then moves the system in the opposing direction, toward stabilization (most systems in nature involve negative feedback loops). These negative feedback loops contribute to homeostasis and the overall balance and functioning of Earth's systems. **Positive feedback loops** have the opposite effect. Increased output leads to increased input, leading to further increased output and destabilization of the system (exponential population growth and Arctic sea ice/glacial melting are two current examples).

▌ Approaching environmental issues in terms of systems is important in understanding how to avoid disrupting the many complex and interrelated Earth processes, and in how to mitigate the disruptions already present. Earth's unperturbed ecosystems use renewable solar energy, recycle nutrients, and are stabilized by negative feedback loops.

### The specific causes and effects of eutrophication

▌ **Eutrophication** is the process of increased nutrient (N and/or P) enrichment → producing algal blooms → followed by eventual microbial decomposition of this increased organic matter → finally resulting in low oxygen levels/hypoxia and the suffocation of aquatic life, including shellfish, fish, and aquatic invertebrates.

▌ Eutrophication of freshwater systems is frequently caused by increased P (in the form of $PO_4$); in marine systems, increased N (in the form of $NO_3$) tends to be the cause.

▌ Hypoxic "dead zones" are increasing, with approximately 500 currently documented worldwide as of 2016. The Gulf of Mexico and the Chesapeake Bay estuary are examples of the most serious cases.

### How to fully define an ecosystem and evaluate the interactions of its living and nonliving entities, especially with regard to productivity

▌ An **ecosystem** consists of all living organisms and the nonliving attributes (temperature, water, sunlight, etc.) that occur and interact in a particular area.

▌ In all ecosystems, *energy flows in a one-way direction* from sun to organic compounds made through photosynthesis by producers, and exiting back to the environment in the form of heat during cellular respiration.

▌ In contrast, *matter is recycled within ecosystems* during feeding relationships involving producers and consumers and ultimately back to the nonliving environment when an organism dies and decays.

▌ **Gross primary productivity (GPP)** is the assimilation of energy into biomass by autotrophs during photosynthesis. A portion of this biomass is used by the autotrophs to power their own metabolism during cellular respiration.

▌ **Net primary productivity (NPP)** is the energy/biomass that remains after respiration (R) and is available for consumption by heterotrophs/consumers.

▌ *NPP = GPP − R*

▌ Ecosystems vary in their **productivity** (the rate at which autotrophs convert energy to biomass). Coral reefs, tropical rainforests, and swamps/marshes have higher productivity than the open ocean and deserts (when comparing $g\,C/m^2/yr$).

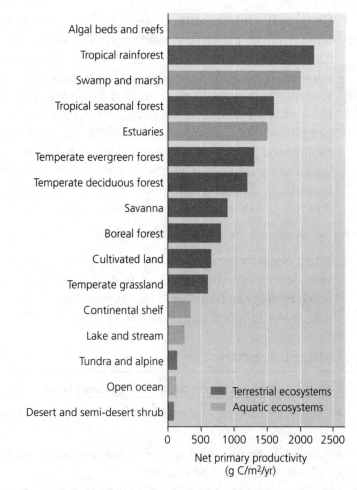

**Figure 5.7**  Net Primary Productivity for Major Ecosystem Types

## *The fundamentals of landscape ecology, GIS, and ecological modeling*

▌ **Landscape ecology** approaches environmental problem solving on a broad, geographic scale that encompasses assessment of the multiple ecosystems involved. This approach allows scientists to assess how landscape structure affects the abundance, distribution, and interaction of organisms. The current Chesapeake Bay restoration of its once-productive oyster industry is an example of a successful landscape ecology approach. It involves a temporary moratorium on oyster harvesting, construction of artificial reefs for young oyster larvae attachment, oyster aquaculture support, and avoidance of the introduction of non-native oysters.

▌ In any given landscape, there are often numerous ecosystem types and **ecotones** (transitional zones) consisting of **patches** spatially arrayed into a **mosaic** of habitats.

▌ **Conservation biology**, the study of the loss, protection, and restoration of biodiversity, must take into consideration all this habitat variety, particularly the fragmentation of habitat into small and isolated patches.

▊ **Geographic information system (GIS)** is a common tool for landscape ecologists, involving computer software that takes multiple types of data (geology, hydrology, vegetation, animal population, and human infrastructure) and combines them on a common set of geographic coordinates in order to more accurately assess environmental problems.

▊ **Ecological modeling** is the practice of constructing and testing models (simplified representations of complex natural systems) that aim to explain and predict how ecological systems function.

▊ Data from published scientific journals are used to establish the model's parameters. These assumptions are then used to predict the effects of different possible approaches to solving the problem.

### *How to assess ecosystem services and their benefit to our lives*

▊ When ecosystems are undisturbed and functioning normally, they provide both goods (lumber, soil, fuel, metals, fiber, etc.) and **ecosystem services** (regulation of oxygen, carbon dioxide, temperature, precipitation, water supplies, etc.) essential to all life.

---

**TABLE 5.1** Ecosystem Services

**Ecological processes do many things that benefit us:**

▶ Regulate oxygen, carbon dioxide, stratospheric ozone, and other atmospheric gases

▶ Regulate temperature and precipitation by means of ocean currents, cloud formation, and so on

▶ Protect against storms, floods, and droughts, mainly by means of vegetation

▶ Store and regulate water supplies in watersheds and aquifers

▶ Prevent soil erosion

▶ Form soil by weathering rock and accumulating organic material

▶ Cycle carbon, nitrogen, phosphorus, sulfur, and other nutrients

▶ Filter waste, remove toxins, recover nutrients, and control pollution

▶ Pollinate plant crops and wild plants so they reproduce

▶ Control crop pests with predators and parasites

▶ Provide habitat for organisms to breed, feed, rest, migrate, and winter

▶ Produce fish, game, crops, nuts, and fruits that people eat

▶ Supply lumber, fuel, metals, fodder, and fiber

▶ Furnish medicines, pets, ornamental plants, and genes for resistance to pathogens and crop pests

▶ Provide recreation such as ecotourism, fishing, hiking, birding, hunting, and kayaking

▶ Provide aesthetic, artistic, educational, spiritual, and scientific amenities

---

▌ One of the most important ecosystem services is the cycling of essential nutrients in biogeochemical cycles.

### *The major biogeochemical cycles (C, N, P, and water), including the source, sink, and importance of each nutrient to all living things*

▌ The main factors of any biogeochemical cycle are the **reservoir** (place where materials are stored), **flux** (rate at which materials move between reservoirs), the **source** (releasing more materials than it accepts), and the **sink** (accepting more materials than it releases).

▌ The **water cycle** affects all other cycles. The ocean is the main reservoir, holding 97% of all Earth's water; however, less than 1% is in a form readily usable (groundwater, surface fresh water, and rain). **Evaporation, transpiration, precipitation**, and surface **runoff** are the main processes determining the availability and location of Earth's water.

▌ The **carbon cycle** circulates carbon, the defining element of all organic compounds (carbohydrates, fats, and proteins). The photosynthetic process of autotrophs takes in $CO_2$ from the atmosphere, synthesizing these organic compounds. Plants are the major living carbon reservoir; however, sedimentary rock/fossil fuels are the largest fossil reservoir. $CO_2$ is returned to the atmosphere by the process of cellular respiration and the burning of fossil fuels.

▌ The **nitrogen cycle** moves nitrogen from the atmosphere through the living world. Nitrogen is an essential element required for making DNA, RNA, proteins, and enzymes.

  ■ The major reservoir of nitrogen ($N_2$) is in the atmosphere (nitrogen constitutes 78% of atmosphere by mass); however, plants and animals cannot utilize this chemically inert form.

  ■ For nitrogen to become biologically available, **nitrogen fixation** ($N_2 \rightarrow NH_3$) must be carried out by symbiotic bacteria living in nodules of legumes (soybeans, peanuts). $NH_3$ then combines with $H_2O$ to form $NH_4^+$ (**ammonification**). Other types of soil bacteria perform **nitrification**, changing $NH_4^+$ to $NO_2^-$ and finally to $NO_3^-$ (nitrates), which can be taken up by plants and subsequently consumed by heterotrophs.

  ■ **Denitrification**, which is carried out by other soil bacteria when organisms die, converts $NO_3^-$ to $N_2$, thereby releasing nitrogen back into the atmosphere and completing the cycle.

▌ The **phosphorus cycle** is the only sedimentary cycle; phosphorus has no appreciable atmospheric reservoir. Because most P is bound up in rock and only slowly released, phosphorus is frequently a limiting factor for plant growth.

*How human activities adversely affect biogeochemical cycles, and possible approaches for improved resource management*

▌ Human activity is greatly affecting the **water cycle** through:

1) deforestation, which intensifies surface runoff and erosion, reduces transpiration, and lowers groundwater tables.
2) the emission of air pollutants, which changes the chemistry of precipitation.
3) severe overdrawing of groundwater for drinking, irrigation, and industrial uses.

▌ Human activity is affecting the **carbon cycle** through the burning of fossil fuels, which releases carbon dioxide and greatly increases the flux of carbon from the lithosphere to the atmosphere. Additionally, by cutting down forests, less vegetation results in less $CO_2$ being extracted from the atmosphere. Overall, the movement of $CO_2$ from atmosphere back to hydrosphere, lithosphere, and biosphere has not kept pace with movement of $CO_2$ into the atmosphere, resulting in the concerns behind today's anthropogenic global climate change.

▌ **Ocean acidification**, the process by which some of the excess $CO_2$ in the atmosphere is being absorbed by ocean water, is an additional concern of increasing atmospheric $CO_2$ levels. The lowering of ocean pH impairs the ability of corals and other shelled organisms to build exoskeletons of calcium carbonate and threatens the existence of coral reefs and many marine invertebrates.

▌ By learning how to synthesize ammonia (**Haber-Bosch process**) on an industrial scale, humans have effectively doubled the rate of nitrogen fixation on Earth, thereby increasing nitrogen's movement from the atmosphere to Earth's surface. When farming practices allow irrigation runoff and soil erosion, nitrogen (in the form of nitrates from synthetic fertilizers produced from the Haber-Bosch process) flows from farms into terrestrial and aquatic ecosystems. This leads to nutrient enrichment/pollution, eutrophication, hypoxia, and a greatly imbalanced **nitrogen cycle**.

▌ Humans are also affecting the natural **phosphorus cycle** by increasing amounts of this limiting factor in fresh water. Increases in phosphorus fertilizer use boosts algal growth, also leading to eutrophication and hypoxic conditions.

▌ Tackling nutrient enrichment ($CO_2$, N, and P) will require diverse approaches because of our current heavy reliance on synthetic fertilizers in modern global agribusiness and on fossil fuels for energy. A *few* examples of possible approaches to control nutrient pollution in the Chesapeake Bay and Mississippi River watersheds are:

1) planting vegetation buffers around streams in the watershed.
2) restoring nutrient-absorbing wetlands along waterways.
3) increasing the practice of sustainable agriculture by reducing fertilizer use on farms and lawns.

4) improving technologies in sewage treatment plants to enhance N and P capture.
5) upgrading stormwater systems to capture runoff from roads and parking lots.

# Chapter 11: Biodiversity and Conservation Biology

**YOU MUST KNOW**

- How to characterize the scope of Earth's biodiversity.
- How to compare the natural background extinction rate with periods of mass extinction.
- How to evaluate the primary causes of biodiversity loss.
- The specific benefits of biodiversity.
- How to analyze the efforts to conserve threatened and endangered species.
- Specific examples of conservation efforts above the species level.

### *How to characterize the scope of Earth's biodiversity*

- To date, of the 3–100 million species currently estimated by scientists, only 1.8 million species have been described and named.
- **Biodiversity** is a multifaceted concept including the variety of life across all levels of biological organization.

  - **Species diversity** focuses on the number or variety of different species in a particular region. **Species richness** (the number of different species) and **species evenness** (the relative abundance of different species) are important components to consider when evaluating species diversity.
  - **Genetic diversity** focuses on the differences in DNA composition between individuals of the same species, which provide the crucial raw material for adaptation to local conditions. Populations with low genetic diversity (cheetahs, bison, elephant seals) are vulnerable to environmental change, disease, and ultimately to extinction.
  - **Ecosystem diversity** refers to the number and variety of ecosystems, biotic communities, or habitats within a specific area. A coral reef contains far more biodiversity than the same area of a monoculture cornfield.

- Some taxonomic groups—for example, insects—show much more species diversity (a high species richness) than others, such as birds and mammals.

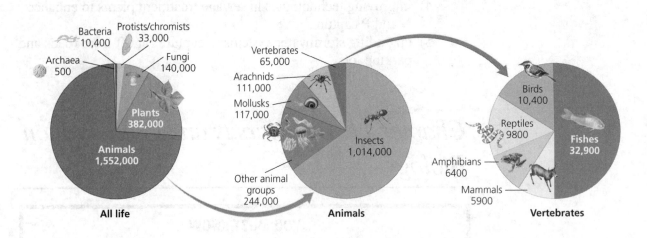

**Figure 11.4** Uneven Species Distribution Among Taxonomic Groups: Insect Dominanace

- Measuring biodiversity is complex. Biodiversity is usually expressed in terms of species richness because this is the easiest component to measure. All estimates of biodiversity are incomplete because of the difficulty of estimating microbes, nematodes, fungi, protists, and soil-dwelling arthropods. Also, many areas of Earth still remain little explored (for example, tropical rainforests and ocean depths).
- Biodiversity is unevenly distributed on Earth; however, species richness generally increases as one approaches the equator, a phenomenon known as the **latitudinal gradient**. This latitudinal gradient influences the species diversity of Earth's biomes, with tropical biomes showing more diversity than high-latitude biomes (boreal forests and tundra).

### *How to compare the natural background extinction rate with periods of mass extinction*

- **Extinction** occurs when the last member of a species dies and the species ceases to exist (e.g., Costa Rica's golden toad and Mauritius's dodo). **Extirpation** occurs when a particular population of a species disappears from a given area, but not from the entire global population (e.g., the Siberian tiger). Over time, extirpation is a process that often leads to extinction.
- Extinction occurs naturally, at a very slow rate over geologic time. Paleontologists estimate that roughly 99% of all species that ever lived are now extinct. **Background rates of extinction** are currently estimated for mammals and marine animals to be 1 species vanishing per year for every 1–10 million species.
- During Earth's five mass extinction episodes, extinction rates rose far above this background rate, often eliminating between 50% and 90% of all species per episode.

- The current (sixth) mass extinction is different from all previous episodes in that humans are now the causal agent, with the current global extinction rate now estimated at *100–1,000 times greater than the background rate.*

- The International Union for Conservation of Nature (IUCN) maintains the **Red List**, an updated list of species facing high risks of extinction. As of 2016, 22% of mammal species, 31% of amphibian species, 16% of fish species, and 15% of all bird species are currently threatened with extinction.

### How to evaluate the primary causes of biodiversity loss

- Four primary causes of population decline and species extinction have been identified by scientists: habitat loss, invasive species, pollution, and overharvesting. Each of these is further intensified by human population growth and increase in per capita consumption of resources.

- Global climate change is currently being considered a fifth major cause.

  - **Habitat loss** is the single greatest cause of biodiversity loss. Habitats are lost by outright destruction, moderate alteration, and, most frequently, **habitat fragmentation**. Farming, logging, and road building intrude into forested habitats, breaking up a continuous forest expanse into an array of fragments or patches.

  - Habitat loss is the primary source of decline for 80% of threatened mammals and birds.

  - Native American prairie biomes have been almost completely converted to agriculture (<1% remain), resulting in the extirpation of 82–99% of grassland birds.

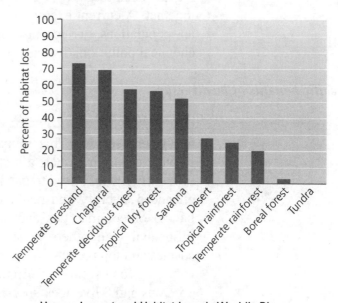

Human Impact and Habitat Loss in World's Biomes

  - Globally, tropical rainforests are experiencing the greatest habitat loss.

■ **Invasive species** are non-native species that thrive in areas where they are introduced, outcompeting, preying on, or otherwise harming native species (examples include the zebra mussel, Asian long-horn beetle, and European starling). Species native to islands are especially vulnerable to invasive species.

■ **Pollution** (air pollution, water pollution, and agricultural pollution from heavy metals, endocrine disrupters, oil, and other chemicals) is a substantial threat, although less significant in its effects than are habitat loss and invasive species. The once-abundant Monarch butterfly has undergone dramatic declines due to herbicides, pesticides, and habitat loss.

■ **Overharvesting** today poses a threat for some top predators such as the K-selected African elephant and Siberian tiger, and also the r-selected shark species. To prohibit overharvesting, many governments have passed endangered species legislation and attempted to reduce **poaching** (the illegal killing of wildlife for meat and body parts).

■ **Global climate change** affects biodiversity everywhere, not just in discrete locations. Organisms are being forced to shift their geographic ranges, and many (especially plants) are unable to move poleward fast enough. In the Arctic, where warming has been greatest, thawing ice is hindering the polar bear's ability to hunt seals.

■ Scientists predict that a rise in global temperature of 1.5–2.5°C could put 20–30% of Earth's biodiversity at risk of extinction.

■ Reasons for the decline of a population or species can be challenging to determine because of the many complex factors working together. A current example is the worldwide collapse of amphibian populations, as recent studies implicate habitat destruction, chemical pollution, disease, invasive species, and climate change.

*The specific benefits of biodiversity*

❚ Biodiversity provides:

■ **ecosystem services** that humans cannot replicate (intact forests provide clean air, water, a hydrologic buffer against flooding and drought) or would need to pay for if nature did not provide them.

■ enhanced stability and resilience (the ability to withstand disturbance) of communities and ecosystems. Some species are more influential than others in an ecosystem's functioning (keystone species and top predators).

■ enhanced food security. Currently we get 90% of our food from just 15 crops and 8 livestock species, as compared to the approximately 7,000 plant species and several thousand animal species in past history. Therefore, maintaining genetic diversity in our existing crops is essential.

- valuable drugs and medicines for humans.
- economic benefits through tourism and recreation, particularly for tropical developing countries that have impressive species diversity.
- an outlet for human's instinctive "love for nature" (**biophilia**, the term developed by Harvard professor emeritus E. O. Wilson).

### *How to analyze the efforts to conserve threatened and endangered species*

- **Conservation biology** is the scientific discipline devoted to understanding the factors that influence the loss, protection, and restoration of biodiversity.
- **The Endangered Species Act (ESA),** passed in 1973, is the prime piece of legislation protecting biodiversity in the United States. This act forbids actions that destroy endangered species or their habitats and also forbids trade in products made from endangered species. The most notable successes of the ESA have been the comeback of the peregrine falcon, brown pelican, and bald eagle following the ban on the pesticide DDT; however, some opponents feel that the ESA values species over jobs (for example, the northern spotted owl versus the old-growth logging industry).
- **The Convention on International Trade in Endangered Species of Wild Fauna and Flora (CITES)** protects endangered species by banning the international transport of organisms or their body parts. The 1990 global ban on the ivory trade is considered the biggest accomplishment of CITES to date, protecting both African elephants and rhinos.
- The **Convention on Biological Diversity** (1992) has 193 signatory nations as of 2010, promoting the conservation of species, using biodiversity in a sustainable manner, and ensuring the fair distribution of biodiversity's benefits.
- Modern recovery strategies include **captive breeding** (California condor) and **reintroduction programs** (gray wolf to Yellowstone National Park). The newest approach involves investigating the potential of **cloning** endangered species.
- **Forensic DNA analysis** is now able to identify a species and its specific geographic region of origin. For example, recent research involving confiscated African elephant ivory has revealed the specific geographic areas where elephants are poached and the extensive crime network and countries involved in threatening this species.

### *Specific examples of conservation efforts above the species level*

- Protecting communities and ecosystems is currently more challenging than protecting individual species because without protecting habitat, protecting endangered and threatened species does little good.
- **Umbrella species** (e.g., the Serengeti's lions and wildebeest) are large species that roam great distances, and are often the focus of current conservation efforts because protecting them translates into protecting many species.

▮ **Parks and protected areas** set aside to remain undeveloped help to conserve habitat, communities, ecosystems, and landscapes. Currently, 13% of Earth's land area is in various types of parks, preserves, and wilderness areas.

▮ **Biodiversity hotspots** have been organized around geographic areas that support a great number of endemic (native/found nowhere else) species. To qualify, an area must have already lost 70% of its habitat as a result of human impact and be at risk of losing more (e.g., Hawaiian Islands, Madagascar, Brazil).

▮ **Ecological restoration** attempts to mitigate the effects of environmental degradation, focusing on reestablishing basic processes (the cycling of nutrients and the flow of energy) that make an ecosystem function, along with bringing back the original population of plants and animals (e.g., Illinois tallgrass prairie, Florida Everglades, southern Iraq's marshlands).

▮ **Community-based conservation** efforts are now being seen as the "key" to protecting global biodiversity because these efforts draw ecotourism, support local economies, and empower residents toward sustainable management of their local environment. East Africa's initiatives with the Masaai and other people of the region are positive examples of community-based conservation attempting to reverse the dramatic declines of endangered African wildlife. Tourist dollars and wildlife management authority have begun to be transferred to local villages and co-managed with Kenya Wildlife Service and international NGOs. It has become clear that in order to conserve animals and ecosystems, the local people need to be stewards and feel invested in conservation.

*Multiple-Choice Questions*

1. Which statement best describes energy transfer in a food web?
   (A) Energy moves from autotrophs to heterotrophs to decomposers, which convert it to usable form, which producers use again.
   (B) Energy is transferred to consumers, which use it to synthesize food.
   (C) Energy from producers is converted to oxygen, thus supporting consumers.
   (D) Energy is transferred to consumers, which convert some to nitrogen, because only nitrogen-fixing bacteria can bring nitrogen into the ecosystem.
   (E) Energy from the sun is stored in green plants and transferred to consumers through herbivory.

2. If the pH of a solution is increased from 7 to 9, it means that the
   (A) concentration of $H^+$ has decreased by 100 times.
   (B) concentration of $H^+$ has increased by 2 times.
   (C) concentration of $OH^-$ has decreased by 10 times.
   (D) concentration of $OH^-$ has increased by 2 times.
   (E) concentration of $H^+$ has increased by 100 times.

3. Given that the half-life of the radioisotope Cs-137 is 30 years, approximately how long will it take for 80 lbs to decay to less than 1 lb?
   (A) 7 years
   (B) 15 years
   (C) 70 years
   (D) 100 years
   (E) 200 years

Questions 4–6. The terms in the key below relate to the processes in the nitrogen cycle. Match a letter from the key with each of the following questions. A letter can be used once, more than once, or not at all.
   (A) nitrogen fixation
   (B) nitrification
   (C) ammonification
   (D) synthetic ammonification (Haber-Bosch process)
   (E) denitrification

4. Which process generates nitrogen gas ($N_2$)?

5. Which process could contribute excess nitrates leading to eutrophication?

6. Which process converts nitrogen gas into a usable form of nitrogen for plants?

7. When environmental conditions change, populations that are *most* vulnerable to extinction are
   (A) populations that have reached their carrying capacity.
   (B) large populations of small individuals living in urban areas.
   (C) large populations of medium-sized individuals living in rural areas.
   (D) populations of species that are generalists with regard to needed resources.
   (E) populations of species that are specialized with regard to needed resources.

8. The climatographs of tundra and desert biomes both
   (A) lack trees.
   (B) have relatively low precipitation on an annual basis.
   (C) lack birds as the dominant consumer.
   (D) have many burrowing rodents.
   (E) have seasonally low temperatures.

9. A trophic cascade is the effect of ___ on ___.
   (A) flooding; terrestrial ecosystems
   (B) detritivores; decomposers
   (C) producers; first-level consumers
   (D) top predators; one another
   (E) tertiary consumers; the abundance of primary consumers

10. Eutrophication processes that have taken place in the Gulf of Mexico and other marine locations are caused primarily by
    (A) an overabundance of algal growth.
    (B) nitrates from agricultural applications.
    (C) pesticides from modern agriculture.
    (D) global warming from human use of fossil fuels.
    (E) low oxygen levels resulting from decomposition.

11. Which of the following is an example of a positive feedback loop?
    (A) an increase in predators, resulting in fewer prey
    (B) the melting of Greenland's glaciers, resulting in sea level rise and continental flooding
    (C) the melting of Arctic sea ice, resulting in increased solar radiation absorption, and more sea ice melting
    (D) an increase in prey, resulting in an increase of decomposers
    (E) human exponential growth

12. A marshland that contains elements of both the forest and a coastal marsh could be called
    (A) an abiotic system.
    (B) an open system, exchanging nutrients.
    (C) an ecotone.
    (D) a closed system, maintaining all nutrients.
    (E) an ecosystem.

13. Which of the following combination of processes can change global species diversity?
    (A) immigration and extirpation
    (B) speciation and extinction
    (C) speciation and immigration
    (D) emigration and extinction
    (E) extirpation and extinction

14. The species most often vulnerable to human impact is the
    (A) keystone species.
    (B) producer.
    (C) flagship species.
    (D) umbrella species.
    (E) top predator.

15. A migratory bat species pollinates cactus plants in northern Mexico on its way to southern Arizona, where it spends the summer eating insects and reproducing. Farmers spraying pesticides affect these bats, which eat the insects and also feed them to their young. This scenario best describes
    (A) a top predator.
    (B) extirpation.
    (C) an umbrella species.
    (D) insect biodiversity loss.
    (E) bioaccumulation.

16. The normal functioning of an ecosystem requires that
    (A) energy be able to change forms, be recycled, and be decomposed.
    (B) energy be able to change forms, given that matter is not able to change forms.
    (C) energy be able to be recycled and matter be able to flow through biogeochemical cycles.
    (D) matter be able to be recycled and energy able to flow through ecosystems.
    (E) matter be able to be recycled, change forms, and be destroyed as a result of ecosystem processes.

17. Assuming that the autotrophs in an ecosystem contain 500,550 kcal of energy, the third trophic level organisms could expect to contain approximately ___ of energy.
    (A) 50,055 kcal
    (B) 5,005 kcal
    (C) 1/3 of 500,550 kcal
    (D) 2/3 of 500,550 kcal
    (E) 500 kcal

18. The process of cellular respiration allows for the normal functioning of ecosystems by
    (A) taking in $CO_2$ and releasing $C_6H_{12}O_6$ and $O_2$ to other organisms.
    (B) taking in $O_2$ and releasing $CO_2$ and $H_2O$ back to the environment.
    (C) taking in $CO_2$ and $H_2O$ and releasing energy and glucose to organisms higher in the food chain.
    (D) taking in $O_2$ and releasing energy and glucose back to the environment.
    (E) taking in $H_2O$ and $CO_2$, producing glucose, and releasing $O_2$.

19. Global climate change could be best thought of as a(n) _____ affecting a population's size.
    (A) logistical growth factor
    (B) exponential growth factor
    (C) density-dependent factor
    (D) density-independent factor
    (E) endemic limiting factor

20. Introduced species can often be described as ___, outcompeting ___ species in a given region.
    (A) specialists; extinct
    (B) endemic; other
    (C) generalists; endemic
    (D) K-strategists; r-strategists
    (E) specialists; extirpated

21. The Hawaiian honeycreeper's specific biotic interactions, including the abiotic factors in its high-altitude forest environment, are collectively referred to as its
    (A) trophic level.
    (B) consumption rate.
    (C) habitat.
    (D) symbiotic placement.
    (E) ecological niche.

22. The energy content and biomass of ___ are highest in any food web.
    (A) top carnivores
    (B) producers
    (C) small carnivores such as spiders and lizards
    (D) detritivores
    (E) decomposers

23. The Everglades Restoration Plan in Florida
    (A) seeks to exterminate numerous invasive fish species and plants.
    (B) will restore natural levels of water flow by undoing numerous damming and drainage projects.
    (C) will result in serious depletion of drinking water supplies in south Florida.
    (D) will probably destroy much of the commercial fishing in the area.
    (E) is a long-term restoration project with the focus of restoring the temperate deciduous forest biome.

24. Individuals of a single species fighting over access to a limited resource is an example of
    (A) character displacement.
    (B) competitive exclusion.
    (C) resource partitioning.
    (D) interspecific competition.
    (E) intraspecific competition.

25. The most productive $(gC/m^2/yr)$ ecosystems in the world are
    (A) swamps and marshes.
    (B) boreal forests.
    (C) the open ocean.
    (D) temperate grasslands.
    (E) temperate deciduous forests.

26. Humans have dramatically altered the movement of nitrogen from
    (A) producers to consumers through increased controlled burning.
    (B) oceans to soils.
    (C) soils to the atmosphere.
    (D) the atmosphere to aquatic systems on Earth's surface.
    (E) deep in Earth to the atmosphere's ozone layer.

27. The rate at which biomass becomes available to consumers is termed
    (A) gross primary production.
    (B) ecosystem productivity.
    (C) cellular respiration.
    (D) secondary productivity.
    (E) net primary productivity.

*Use graph below to answer questions 28 and 29.*

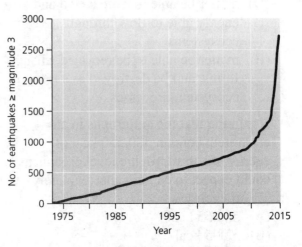

Earthquake Frequency in South Central U.S., 1973-2015

28. The hypothesis that scientists proposed for the results shown in the above graph was
    (A) decreased fault stability.
    (B) increased coal removal.
    (C) decreased environmental protection.
    (D) increased wastewater well injection.
    (E) increased groundwater removal.

29. The approximate percent change in earthquake frequency from 1995 to 2015 was
    (A) 4.5%.
    (B) 22.5 %.
    (C) 225%.
    (D) 450%.
    (E) 500%.

30. A population of grazing, deerlike mammals is found on a remote island in forested New Guinea. Which of the following information is most important in determining whether these mammals all belong to a single species?
    (A) that they share 99% of their physical traits
    (B) whether they can breed with one another
    (C) that the males all have similar antlers
    (D) whether the matings produce viable, fertile young
    (E) that all these deerlike mammals appear to eat the same varieties of grass

31. Plate tectonics is a process that can affect
    (A) the shape of Earth's landscapes, including building mountains.
    (B) normal ecosystem structure and function.
    (C) flow of matter and recycling of energy.
    (D) the frequency of volcanic eruptions coming from tsunami events.
    (E) the number of trophic levels of ecosystems near plate boundaries.

32. Climate change researchers have recently been asking whether forest trees will sequester more carbon as $CO_2$ levels rise, whether this will change as the trees grow, and what is the role of tropospheric ozone's interaction with $CO_2$. Currently, their findings indicate
    (A) there is no interaction between $CO_2$ and tropospheric ozone.
    (B) increasing levels of $CO_2$ always promote increasing plant growth.
    (C) increasing levels of $CO_2$ do not promote further plant growth.
    (D) increasing levels of $CO_2$ and tropospheric ozone discourage the effects on forests by disease and pests.
    (E) elevated $CO_2$ levels increase photosynthesis and tree growth, but moderate levels of ozone offset this increased growth.

33. Assume that there are 1000 mice in a field. Over the course of a year, this population has 200 births, 100 deaths, 100 immigrants, and 100 emigrants. What is the growth rate for this population?
    (A) 100%
    (B) 80%
    (C) 50%
    (D) 10%
    (E) 1%

34. A population's growth rate is most accurately determined by knowing its
    (A) crude birth and crude death rates multiplied by 100%.
    (B) limiting factors.
    (C) crude birth rates, crude death rates, and migration rates.
    (D) survivorship curve type.
    (E) specific density.

35. Zooplankton populations in Lake Erie and the Hudson River have declined by up to 70% since the arrival of zebra mussels because
    (A) waste from zebra mussels promotes bacterial growth that kills zooplankton.
    (B) zebra mussels prey on zooplankton.
    (C) zebra mussels carry a parasite that kills zooplankton.
    (D) zebra mussels feed on cyanobacteria, which zooplankton need as a food resource.
    (E) zebra mussels block sunlight penetration into lakes and thus prevent zooplankton from photosynthesizing.

36. In a New England White Pine forest, the gross primary productivity (GPP) is 70 grams per square meter per day. Respiration loss is 30%. What is the net primary productivity for this forest?
    (A) 10 g/m$^2$/day
    (B) 21 g/m$^2$/day
    (C) 40 g/m$^2$/day
    (D) 49 g/m$^2$/day
    (E) 121 g/m$^2$/day

37. A keystone species
    (A) is most often dominant in numbers in their ecosystem.
    (B) when removed, causes the removal of a trophic level in the ecosystem.
    (C) is most likely a top predator.
    (D) is most likely to be a producer on which all species ultimately depend.
    (E) feeds upon a variety of species in several different trophic levels.

38. Currently, one of the most serious human alterations of the water cycle is
    (A) due to deforestation and resulting erosion.
    (B) due to overdrawing groundwater for irrigation, industry, and drinking water.
    (C) due to damming waterways and the resultant disturbance to fish migrations.
    (D) due to burning fossil fuels and the resultant acid precipitation.
    (E) due to increased desalinization plants and the resultant destruction of marine life.

39. The largest reservoir of carbon in the carbon cycle is
    (A) freshwater systems.
    (B) plant life.
    (C) sedimentary rock and fossil fuels.
    (D) located in the atmosphere.
    (E) located in the hydrosphere.

40. "Dead zones" characteristic of the Chesapeake Bay estuary and many other locations worldwide most frequently result from
    (A) negative feedback loops.
    (B) invasive species of fish displacing native oysters.
    (C) a disrupted water cycle.
    (D) a disrupted nitrogen cycle.
    (E) a disrupted carbon cycle.

41. Ocean acidification results from
    (A) increased use of artificial fertilizers.
    (B) increased pH due to high levels of carbonic acid.
    (C) lowered pH due to carbonic acid.
    (D) the increased availability of carbonate ions.
    (E) the dissolving of shelled organisms.

42. Earth's current biodiversity has most likely resulted from
    (A) high rates of mutation in the past millennium.
    (B) sympatric speciation.
    (C) current climate change pressures.
    (D) speciation minus extinction.
    (E) the wide variety of habitats to fill.

43. The main reason for the high species extinction rate that we see today is
    (A) global climate change.
    (B) invasive disease organisms.
    (C) overharvesting of Earth's valuable biodiversity.
    (D) the inability of Earth's biodiversity to adapt.
    (E) habitat destruction.

44. The level of ecological organization described by ecologists when studying multiple interacting species that live in the same area is
    (A) the habitat.
    (B) energy flow.
    (C) the population.
    (D) the community.
    (E) the ecosystem.

*Free-Response Question*

At Hakalau Forest National Wildlife Refuge on the island of Hawaii, biologists have been working for over 20 years to understand the population change of native forest birds. Much of the region's native forests had been cleared for cattle ranching years earlier, and numerous introduced species have begun to thrive. In 1987, field biologists divided the refuge into 15 transects, with approximately 22 observation points along each transect. A standard method for estimating bird populations (i.e., "point counts" performed by stopping for an 8-minute bird count at each of the 22 observation points) was carried out, with data collection continuing until 2008. Population density data for the Hawaiian 'elepaio, a native forest bird, typify data gathered at Hakalau Forest during this time. Researchers used a statistical method to determine a "line of best fit" that most accurately represents trends in the data. The graph's longer line shows the trend over the entire study, and the shorter graph line shows the trend over the most recent 9 years of the study.

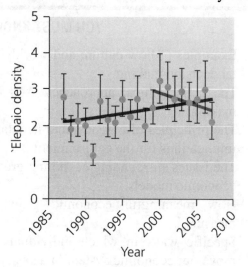

Population Density Data
for the Hawaiian 'elepaio,
1987–2008

a) Like the endangered 'elepaio, fully explain three *species characteristics* that would make other species particularly vulnerable to extinction.

b) Calculate the percent change of the Hawaiian 'elepaio density between 1991 and 2001.

c) Other than initial population size and population density, describe one additional characteristic of a population that ecologists must consider before being able to predict the future dynamics of the population.

d) Explain two *possible factors* causing the density trend in the last 9 years of the study.

e) Describe three strengths of the experimental design for this field study at Hakalau Forest National Wildlife Refuge in Hawaii.

# Environmental Decision Making

## Chapter 6: Ethics, Economics, and Sustainable Development

---

**YOU MUST KNOW**

- Examples of how culture and worldview influence environmental decision making.
- The major ethical approaches used in environmental decision making.
- The principles of classical and neoclassical economics and their implications for the environment.
- The issues surrounding economic growth and "limits to growth" economic models.
- How meaningful economic growth and progress can be measured.
- Specific ways in which individuals and businesses can help move our economic system in a sustainable direction.
- Examples of how sustainable development is a growing global movement.

---

### *Examples of how culture and worldview influence environmental decision making*

- People with different worldviews can study the same situation, yet draw dramatically different conclusions.
- Religion and spiritual beliefs are among the most influential. In addition, political ideology, one's view on the proper role of government, and the level of scientific understanding in the culture play critical roles in environmental decision making.

▮ For example, two Costa Rican ranchers owning identical landholdings may make different management decisions: One might choose to receive government payments and conserve the forests on the land, whereas the other might opt to clear all the land for cattle grazing.

▮ Debates pitting economic arguments for harvesting forests (and other natural resources) versus ethical and environmental arguments for the preservation and restoration of forested lands (as is being actively practiced by Costa Rica's *Payment for Environmental Services* [PSA] program) are occurring in numerous places on Earth today.

### *The major ethical approaches used in environmental decision making*

▮ **Ethical standards** are influenced by the values and moral principles held by a person or culture and are the criteria that help differentiate right from wrong.

▮ Ethics and economics give us tools needed to pursue the triple bottom line: environmental, economic, and social sustainability.

▮ The value of something (a natural resource, such as clean water, or an animal, such as the whooping crane) can be either **utilitarian** (i.e., it is valuable because it is *useful or beneficial* in some way), **intrinsic** (i.e., it is valuable *in itself*), or both.

▮ **Environmental ethics** is the application of ethical standards to relationships between people and nonhuman entities and involves issues of sustainability, environmental justice, and intrinsic values.

▮ An **anthropocentric** ethical view is one that is human-centered and often ignores the notion that nonhuman entities also have intrinsic value. An anthropocentrist evaluates the costs and benefits of actions solely according to their impact on people.

▮ A **biocentric** ethical view ascribes intrinsic value to all living things, including nonhuman life. A biocentrist might oppose an action if it would destroy or degrade plants and animals in an area, even if it would create jobs and generate economic growth.

▮ The **ecocentric** ethical view assesses actions in terms of their benefit or harm to the integrity of whole ecological systems, both living and nonliving. An ecocentrist might oppose an action after broadly assessing the potential effects on water quality, air quality, wildlife populations, and ecosystem services. This approach seeks to preserve the connections vital to the many complex ecosystem entities.

▮ The **preservation ethic** (preserving natural systems intact and for their own intrinsic value) was pioneered by **John Muir** in the early 20th century.

▮ The **conservation ethic** (advocating wise use of natural resources—a utilitarian standard) was pioneered by **Gifford Pinchot**, the first chief of the U.S. Forest Service.

▌ The damming of Hetch Hetchy Valley in Yosemite National Park, to provide drinking water for San Francisco in the early 1900s, was an early example of Pinchot's conservation ethic succeeding over Muir's preservation ethic.

▌ **Aldo Leopold's** "land ethic" called upon humans to include the environment in their ethical framework. By the mid-20th century, this framework had evolved toward an ecocentric outlook.

▌ In the 21st century, the concept of **environmental justice** has emerged, positing the importance of the fair and equitable treatment of all people with respect to environmental policy and practice, regardless of their income, race, or ethnicity.

　　■ Critics have characterized the attempts of Latino farm workers in California's San Joaquin Valley to achieve cleaner air (due to industrial agriculture's pesticide and feedlot emissions and windblown dust from eroding farmland) as important environmental justice issues.

　　■ Uranium mining on Native American Navajo lands in the United States has also raised many of the same environmental justice issues.

　　■ The dumping of hazardous wastes by wealthy countries in poorer countries represents the international extent of environmental justice concerns.

### The principles of classical and neoclassical economics and their implications for the environment

▌ **Economics** is the study of how people decide to use potentially scarce resources to provide goods and services in the face of demand for them.

▌ The **economy** (a social system that converts resources into **goods** and **services**) exists within the environment. The material inputs and the waste-absorbing capacity that Earth can provide to economies are ultimately finite; however, traditional economic schools of thought have viewed environmental resources as free and limitless, and wastes as something that can be endlessly exported and absorbed by the environment at no cost.

▌ The relationships among economic and environmental conditions are becoming widely recognized in the 21st century. In 2005, scientists with the Millennium Ecosystem Assessment concluded that 15 of 24 ecosystem services surveyed globally were being degraded or used unsustainably, and that this degradation could easily disrupt national and global economies.

▌ **Classical economics**, a philosophy founded by Adam Smith, argues that self-interested economic behavior will benefit society as long as the behavior is constrained by law and private property rights, and is operating within a competitive marketplace.

- **Neoclassical economics** includes the psychological factors underlying consumer choices and considers supply, demand, costs, and benefits. Today's market systems operate largely in accord with the principles of neoclassical economics.
- **Cost-benefit analysis**, an approach of neoclassical economics, is used to evaluate public projects, assessing the capital costs that must be paid in order to gain benefits/value for a large group of people. If benefits exceed costs, the action goes forward; if costs exceed benefits, the project is not pursued. A major problem with this approach has been that monetary benefits are usually more easily quantified than environmental costs and benefits.
- Neoclassical economic models involve several assumptions that have had negative implications for the environment.

   - Resources are infinite or substitutable; however, we live on a finite earth with finite resources.
   - Costs and benefits are **internal** (experienced by the buyer and seller alone); however, many transactions affect other members of society (those living "downstream" of the activity).
   - Future effects should be discounted; however, many environmental problems unfold *gradually* (resource depletion, pollution increase).
   - Growth is good; however, sociologists have coined a word for the way that consumption and material affluence often fail to bring contentment: affluenza.

### The issues surrounding economic growth and "limits to growth" economic models

- **Economic growth** is defined as an increase in an economy's production and consumption of goods and services.
- Today's mainstream economic theory assumes that growth is good and can continue forever; however, all matter and nonrenewable resources are finite.
- Today's mainstream economic theory often does not take into consideration **externalities** (costs affecting members of society other than the immediate buyer or seller). By ignoring these costs, economies often create a false idea of the true cost of a particular choice; for example, the air and water pollution generated from building and operating a factory are externalities.
- As human population and consumption continue to grow, it has become clear that natural resources are finite and that growth can continue only when better technologies and approaches lead to improvements in *efficiency* of production.

▌ The study by Meadows, Randers, and Meadows, *Limits to Growth*: *The 30-year Update* (2004), shows that with policies aimed at sustainability (reducing human population growth, making resource use much more efficient, and placing greater emphasis on reusing and recycling resources), we can move towards a **steady-state economy** (an economy that neither grows nor shrinks, but actually mirrors natural ecological systems).

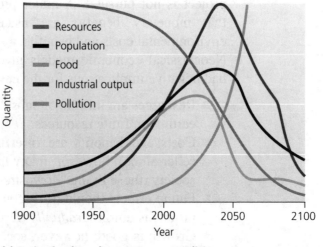

**(a) Projection based on status quo policies**

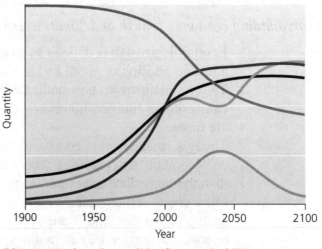

**(b) Projection based on policies for sustainability**

**Figure 6.12a & b**  Limits to Growth Modeling Projections

▌ Every population in nature has a carrying capacity. **Ecological economics** maintains that civilizations, like natural populations, cannot permanently overcome their limits to growth and should not expect endless economic growth, but should support the principles of stability, steady-state, and sustainability. This approach mirrors natural ecological systems.

*How meaningful economic growth and progress can be measured*

▐ **Gross Domestic Product (GDP)** is the total monetary value of final goods and services a country produces each year.

  ▪ Because it doesn't account for **nonmarket values** (values not usually included in the price of a good or service such as ecosystem services), GDP is considered a poor measure of economic well-being.

  ▪ GDP expresses all economic activity, both desirable and undesirable economic activity.

  ▪ Pollution generally causes GDP to increase twice: once when a polluting substance is manufactured, and once when society pays for the cleanup.

▐ **Genuine Progress Indicator (GPI)** differentiates between desirable and undesirable economic activity and is a more accurate indicator of a society's well-being. When compared with GDP, which has risen greatly, the GPI has remained largely flat for the past 35 years.

▐ Because the benefits of ecosystem goods and services lie chiefly in their nonmarket, intrinsic value, economists have historically failed to account for them.

▐ In a major study (*Nature*, 1997), Robert Costanza estimated the value of the world's ecosystem services at more than $33 trillion ($143 trillion in 2014 dollars).

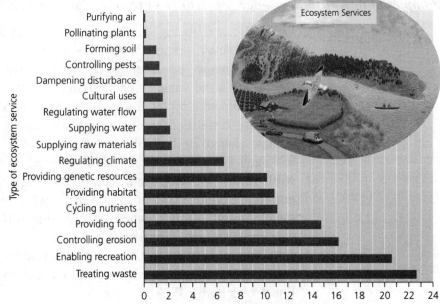

**Figure 6.14**  The Value of Earth's Ecosystem Services

***Specific ways in which individuals and businesses can help move our economic system in a sustainable direction.***

▌ Consumer choice tied with improved environmental knowledge in the marketplace encourages businesses to pursue sustainable practices.

▌ **Ecolabeling** provides businesses a powerful incentive to move towards more sustainable processes when these products are preferentially purchased. Examples include labeling tuna as "dolphin-safe," labeling recycled paper, and labeling organically grown food.

▌ Many corporations (IBM, Hewlett-Packard, Nike, Inc., The Gap, Inc.) are responding to sustainability concerns and modifying their operations through a variety of measures: using high levels of recycled materials in their manufacturing of new products, running alternative transportation programs for employees, increasing energy efficiency of their plants, reducing toxic emissions (including greenhouse gas emissions), and building green corporate headquarters using recycled materials, solar panels, green roofs, and enhanced insulation.

▌ Pearson Education, publisher of the Withgott/Laposata *Environment: The Science behind the Stories* textbook, is the first global media company to become climate-neutral, eliminating its net carbon emissions by installing solar and wind energy to power its facilities. Additionally, Pearson funds forest conservation efforts, has reduced its use of plastic and styrofoam by 85% and 50% respectively, and has been recognized by the Dow Jones Sustainability World Index for over a decade.

***Examples of how sustainable development is a growing global movement***

▌ Economic well-being and environmental well-being are compatible goals; however, some adjustments to the concept of limitless growth must be made by using alternative ways of measuring growth. Recognizing the value of ecosystem services will play an essential role in environmental protection and conservation of resources.

▌ Monetary benefits of reducing greenhouse gas emissions to minimize climate change in the United States have been estimated to be between $235 and $334 billion (U.S. Environmental Protection Agency, 2015).

▌ **Millennium Development Goals for 2015** were set forth in 2000, describing eight broad goals for sustainable development. **Sustainable development** occurs where three sets of goals overlap: social, economic, and environmental.

▌ Programs that pay for ecosystem services are one example of a sustainable development approach that seeks to satisfy a triple bottom line. Costa Rica's PSA program aims to enhance its citizens' well-being by conserving the country's natural resources, while compensating affected landholders for any economic losses.

▌ Valuation of ecosystem services, ecolabeling, ecotourism, corporate sustainability, and alternative means of measuring growth are all helping to bring economic and ethical approaches to bear on environmental protection and sustainable development.

▌ Growing evidence of climate change effects (Hurricane Sandy, flooding in Miami, drought in California) is providing the needed impetus to address climate change and true sustainable development.

▌ Proponents of *strong sustainability* value the importance of natural capital (ecosystem services), insisting that human-made capital alone cannot be the only measure of economic progress, growth, and sustainability.

# Chapter 7: Environmental Policy: Making Decisions and Solving Problems

---

**YOU MUST KNOW**

- The nature of environmental policy and its societal context.
- The role of science in creating environmental policy.
- The institutions important to U.S. environmental policy.
- A brief historical context for U.S. environmental policy, including several major U.S. environmental laws.
- Specific approaches to creating environmental policy.
- The major institutions involved with international environmental policy, including examples of major international treaties.

---

### The nature of environmental policy and its societal context

▌ **Policy** is a tool for decision making and problem solving that makes use of information from science as well as values from ethics and economics.

▌ **Environmental policy** pertains to human interactions with the environment. Its aims are to promote human welfare and to protect natural systems by regulating resource use and pollution effects.

▌ Effective policy requires input from science, ethics, and economics. The recent **hydraulic fracturing** (the process of extracting gas by deep-drilling of shale formations, followed by pumping water, sand, and chemicals under great pressure to fracture the shale, releasing gas) **of the Marcellus Shale** by Cabot Oil and Gas Corporation in Dimock, PA, demonstrates the need for scientific analysis and effective environmental policy because:

- Fracking has been free from many regulatory constraints to encourage replacement of coal use with cleaner-burning natural gas, also increasing local employment and lowering gas prices.
- There has been exemption from seven major federal environmental laws, including NEPA and the Safe Drinking Water Act.
- The chemical additives used in the fracking process are not required to be reported. Wastewater resulting from fracking has been shown to contain radioisotopes and methane gas.
- In 2015, the first scientific study to find clear evidence of aquifer pollution by fracking in the PA Marcellus Shale deposits was published.
- Air pollution (focusing on the powerful greenhouse gas methane) from fracking is also receiving increased attention from researchers.

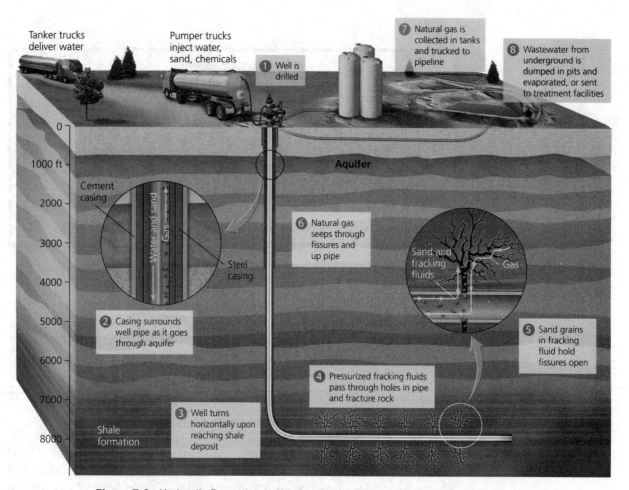

**Figure 7.2** Hydraulic Fracturing Is Used to Extract Natural Gas from Shale Deposits

- Fracking has spread from Pennsylvania to California, and citizens are experiencing impacts on air, water, and land. Some regions have welcomed the gas industry, exempting them from regulations; however, other states have completely banned fracking (NY, VT, and MD).

- Policy is often required to protect resources held and used in common by the public; otherwise, a "**tragedy of the commons**" scenario may result. When publicly accessible resources are open to unregulated use (often because no one person, group, or nation "owns" the resource), these resources tend to become overused, damaged, or depleted. Garrett Hardin first articulated this important concept in his 1968 essay, "Tragedy of the Commons." As a result of understanding this aspect of human behavior, restriction of use and some type of management (regulation, privatization, or cooperative efforts) are now central to environmental policy.

- Effective environmental policy also addresses **external costs** (so that the people or the environment outside the transaction do not bear the harmful impacts created by the transaction) and **environmental justice** considerations.

- Factors hindering effective environmental policy: political lobbying, campaign contributions, and vested interests.

### *The role of science in creating environmental policy*

- Creating environmental policy is a multi-step and time-intensive process involving many people and groups.
- Several of the essential first steps involve the critical role played by science.

  - **Identifying the problem:** The environmental effects of hydro-fracking in Dimock, PA, required an understanding of many science problems including water flow dynamics, the ecological and health impacts of water chemistry, and the ability to adequately and reliably select which citizen wells should be tested.
  - **Identifying the specific causes of the problem:** Much of this scientific research takes place in the arena of risk assessment.
  - **Proposing feasible solutions to the problem:** Much of this scientific research takes place in the arena of risk management in which scientists develop strategies to minimize risk.

- Sometimes science can be "politicized" when policymakers choose to ignore scientific data and allow political ideology alone to determine policy. The scientific community recently claimed that this is occurring in the areas of climate change policy and endangered species protection.

### *The institutions important to U.S. environmental policy*

▌ Environmental law, or legislation, is created by Congress. Once laws are enacted, their implementation and enforcement is assigned to the appropriate executive agency. A few of the most environmentally important are listed below.

   ▪ Department of the Interior—National Park Service, Fish & Wildlife Service
   ▪ Department of Agriculture—National Forest Service
   ▪ Department of Commerce—National Marine Fisheries Service
   ▪ Department of Health and Human Services—Food and Drug Administration, FDA

**TABLE 7.1  Federal Administrative Agencies That Influence Environmental Policy**

| |
|---|
| **EXECUTIVE OFFICE OF THE PRESIDENT** |
| Council on Environmental Quality |
| **DEPARTMENT OF AGRICULTURE (USDA)** |
| Natural Resources Conservation Service |
| U.S. Forest Service (USFS) |
| **DEPARTMENT OF COMMERCE** |
| Bureau of the Census |
| National Marine Fisheries Service |
| National Oceanic and Atmospheric Administration (NOAA) |
| **DEPARTMENT OF DEFENSE** |
| Army Corps of Engineers |
| **DEPARTMENT OF ENERGY (DOE)** |
| Energy Efficiency and Renewable Energy |
| Energy Information Administration (EIA) |
| Federal Energy Regulatory Commission (FERC) |
| **DEPARTMENT OF HEALTH AND HUMAN SERVICES** |
| Centers for Disease Control and Prevention (CDC) |
| Food and Drug Administration (FDA) |
| **DEPARTMENT OF THE INTERIOR** |
| Bureau of Indian Affairs |
| Bureau of Land Management (BLM) |
| Bureau of Reclamation |
| Minerals Management Service |
| National Park Service (NPS) |
| U.S. Fish and Wildlife Service (USFWS) |
| U.S. Geological Survey (USGS) |
| **DEPARTMENT OF LABOR** |
| Occupational Safety and Health Administration (OSHA) |
| **INDEPENDENT AGENCIES** |
| Consumer Product Safety Commission |
| Environmental Protection Agency (EPA) |
| National Aeronautics and Space Administration (NASA) |
| Nuclear Regulatory Commission (NRC) |

*Source:* U.S. General Services Administration, Washington, D.C.

▌ State and local governments also make environmental policy; however, if state or local policies conflict with federal policies, federal laws take precedence.

▌ The Fifth Amendment ensures that private property shall not "be taken for public use without just compensation." A **regulatory taking** occurs when the government, by means of a law or regulation, deprives a property owner of some economic use of his or her private property.

## *A brief historical context for U.S. environmental policy, including several major U.S. environmental laws*

▌ From 1780s to late 1800s: U.S. policy reflected the perception that Western lands were practically infinite and inexhaustible in natural resources. All policies encouraged settlers and land development (Homestead Act of 1862).

▌ From late 1800s to mid-20th century: As the U.S. continent became more populated and its resources increasingly exploited, public policies began to reflect elements of conservation, preservation, and mitigation of early land use (National Park Service Act, Soil Conservation Act, Wilderness Act).

▌ From mid-20th century to 2000: With continued development and industrialization, this era of environmental policy targeted pollution and has given us many of today's major environmental laws.

   ▪ **National Environmental Policy Act (NEPA)**, 1970: Often described as the cornerstone of U.S. environmental law, NEPA established the Council on Environmental Quality and the Environmental Protection Agency (EPA), and it required that an **environmental impact statement (EIS)** be prepared for any federal action (building an interstate highway or dam; logging) that might affect environmental quality.

   ▪ **Clean Water Act**, 1972: Established goal of creating "fishable, swimmable" waterways by setting maximum permissible amounts of pollutants that can be discharged into waterways and requiring permits for the discharge of these pollutants. Recent amendments have focused on stormwater pollution issues, requiring industrial stormwater discharges and municipal sewage discharge facilities to acquire permits prior to operation.

   ▪ **Clean Air Act**, 1970: Established air quality standards for primary and secondary air pollutants and required states to develop specific plans for cleaner air, including the emission testing of automobiles.

   ▪ **Endangered Species Act,** 1973: Identifies threatened and endangered species in the United States, puts their protection ahead of economic consideration, protects their habitat, and directs the Fish & Wildlife Service to prepare effective recovery plans.

   ▪ **Comprehensive Environmental Response, Compensation, and Liability Act (CERCLA),** 1980: Established the Superfund for emergency response and remediation of toxic and hazardous waste sites.

- **Resources Conservation and Recovery Act (RCRA)**, 1976: The primary law pertaining to the disposal of solid and hazardous wastes, setting standards and requiring permits for "cradle to grave" management.
- **Safe Drinking Water Act**, 1974: Authorizes the EPA to set quality standards for tap water provided by public water systems and to protect drinking water from sources of contamination.

- The **Environmental Protection Agency** (EPA) has been the main U.S. agency working to develop solutions to pollution issues, conducting and evaluating environmental research, setting and enforcing standards for pollution levels, and monitoring overall environmental quality.
- **Sustainability** (safeguarding functioning ecosystems while raising living standards for the world's developing nations) and **climate change** have been the focus of 21st-century U.S. environmental policy.

### Specific approaches to creating environmental policy

- Environmental policy is diverse and follows three major approaches that include lawsuits in the courts, command-and control policy, and economic policy tools.
- **Congressional legislation** and subsequent regulations from the appropriate administrative agencies, such as the Department of the Interior and the Department of Agriculture, comprise the command-and-control approach in which the regulating agency prohibits certain actions, sets limits or standards, and threatens punishment for those violating these terms.
- Shortcomings of this top-down approach have led many economists to advocate alternative economic policy tools such as:

  - **Green taxes**—The polluter-pays principle, which levies taxes on businesses that engage in environmentally harmful activities and products, provides a market-based incentive to *discourage*/correct the activity.
  - **Subsidies**—The government provides an incentive (cash or tax break) intended to *encourage* a particular industry or activity. Subsidies can be used to promote environmentally sustainable activities, but they can also be used to support unsustainable ones such as fossil fuel corporations. Most of the subsidies for renewable energy go toward corn ethanol, which is not widely viewed as a sustainable fuel.
  - **Ecolabeling**—Sellers who use sustainable practices advertise this fact, hoping to win approval and sales from buyers.
  - **Permit trading**—The government issues marketable emissions permits to businesses that conduct environmentally harmful activities; the businesses then engage in buying and selling these permits to each other. Under a **cap-and-trade system**, an acceptable level of pollution is determined by the government,

which then issues permits. A company receives credit for amounts it does not emit and then can sell this credit to other companies unable to meet the capped standard. The U.S. has employed the cap-and-trade system for dealing with the stricter $SO_2$ standards of the 1990 Clean Air Act amendments.

- **Market incentives** operate at the local level. Examples would be 1) **charging** for waste disposal according to amount generated, and 2) issuing **rebates** to residents who buy environmentally efficient appliances, such as water-efficient toilets.

### *The major institutions involved with international environmental policy, including examples of major international treaties*

- The **United Nations (UN)**, today serving every nation in the world, has taken an active role in shaping international environmental policy.
- The **World Bank**, based in Washington, D.C., is one of the largest sources of funding for economic development worldwide, providing funding for irrigation infrastructure, dams, and other major development projects. Some World Bank projects have been criticized as unsustainable, often causing more environmental problems than they solve.
- The **European Union's (EU)** Environment Agency works to address waste management, noise, pollution of water and air, and habitat degradation for its 27 member nations.
- The **World Trade Organization (WTO)** imposes financial penalties on those nations that do not comply with its directives. Thus, the WTO can exercise authority in a nation's internal affairs. The UN and the EU, in contrast, do not exercise authority within the borders of their member nations.
- **Nongovernmental organizations (NGOs)** are becoming increasingly influential in their environmental protection advocacy role. The Nature Conservancy, Greenpeace, and Conservation International are just a few of these influential NGOs.
- Some major international environmental treaties:

    - **Convention on International Trade in Endangered Species of Wild Fauna and Flora (CITES),** 1975: Protects endangered species by banning their commercial trade as live specimens or wildlife products.
    - **Montreal Protocol,** 1989: This treaty's goal is to reduce the emission of airborne chemicals (CFCs) that deplete stratospheric ozone. Because of its effectiveness in decreasing global CFC emissions, it is considered the most successful effort to date in addressing a global environmental problem.
    - **Kyoto Protocol,** 2005: An agreement drafted in 1997 that calls for reducing, by 2012, emissions of 6 greenhouse gases to levels lower than their 1990 levels. The United States has refused to ratify this treaty, primarily on the basis of economics.

## *Multiple-Choice Questions*

1. Which of the following is an assumption of modern, 21st-century economics?
   (A) Economic growth is good.
   (B) Resources are infinite.
   (C) Internal costs and benefits are valuable and always considered.
   (D) All citizens deserve just and equal treatment.
   (E) An anthropocentric worldview is predominant.

2. The National Environmental Policy Act (NEPA)
   (A) puts all federal land under environmental oversight and protection.
   (B) requires compensation be given to anyone harmed by pollution from a business or corporate entity.
   (C) requires environmental impact statements for any federal project.
   (D) protects all endangered species from international trade.
   (E) was recently signed into law by President Barack Obama.

3. A(n) ___ is best defined as one who evaluates an action based on its impact on water and air quality.
   (A) biocentrist
   (B) ecocentrist
   (C) anthropocentrist
   (D) ethnocentrist
   (E) modern market economist

4. A cap-and-trade system
   (A) allows industries to set their own levels for pollution emissions so that trading can continue.
   (B) has recently been repealed by Congress as useless in diminishing pollution.
   (C) specifies a certain cap on industrial pollutants that can be traded to other nations.
   (D) rapidly brings pollution emissions to near zero for participating industries.

   (E) permits industries that pollute under the federal limit to sell credits to industries that pollute over the limit.

5. A cost-benefit analysis for a public project always includes
   (A) the internal costs.
   (B) the external costs.
   (C) the GDP.
   (D) the GNP.
   (E) the intrinsic value.

6. Regarding environmental laws and the U.S. Constitution,
   (A) state laws take precedence over federal laws.
   (B) federal laws can override the Constitution if a waiver fee is paid.
   (C) fundamental environmental policy is always set forth by the Constitution.
   (D) federal environmental legislation always takes precedence over state laws.
   (E) environmental decision making always takes place on a case-by-case basis, with the most appropriate decision taking precedence.

7. A modern-day example of the "tragedy of the commons" would be
   (A) strip-mining in a national park or national forest.
   (B) use of herbicides and pesticides by industrial agriculture in California.
   (C) acid precipitation falling throughout the northeastern United States.
   (D) grazing sheep in a private park.
   (E) overfishing in a privately owned lake, greatly reducing its biodiversity.

8. Hydraulic fracturing is the process used to
   (A) obtain new resources of shale and coal.
   (B) lessen our dependence on fossil fuels.
   (C) increase our natural gas supplies.

(D) fracture deep shale deposits with acidic chemicals.

(E) promote renewable energy reserves.

9. External costs include
(A) an employee's wages.
(B) energy costs to operate a factory.
(C) acid rain damage caused by factory.
(D) worker's compensation and insurance coverages.
(E) raw material extraction to operate the factory.

10. One of the most effective international environmental treaties to date is the
(A) Kyoto Protocol.
(B) Montreal Protocol.
(C) CITES.
(D) NEPA.
(E) ESA.

11. Which of the following accounts for the nonmarket value of goods and a country's overall desirable economic activity?
(A) GDP
(B) GAP
(C) GPI
(D) GNP
(E) steady-state growth

12. Modern environmental policy aims to
(A) promote fairness among people and groups in the use of resources.
(B) use natural resources for economically important industrial products.
(C) protect the values of the legal landowner.
(D) preserve all present natural areas in the most pristine condition possible.
(E) promote economic growth whenever possible.

13. The economic value of ecosystem services was studied and first estimated in the late 20th century by
(A) Aldo Leopold.
(B) Robert Costanza.

(C) John Muir.
(D) Gifford Pinchot.
(E) Rachel Carson.

14. Once an environmental law or regulation having to do with the timber industry has been passed, it is up to the ___ to carry out the enforcement of this legislation.
(A) Environmental Protection Agency
(B) Department of the Interior
(C) Department of Agriculture
(D) Department of Health and Human Services
(E) National Environmental Policy Act

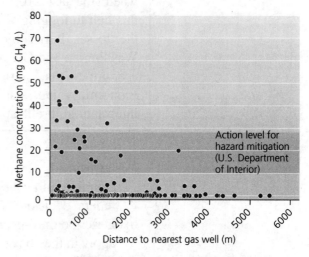

Methane concentration and distance to gas wells

15. The above figure best demonstrates that
(A) fracking is associated with increased seismic activity.
(B) methane results from fracking.
(C) methane concentration increases with fracking.
(D) methane concentrations have an inverse relationship to drinking water well proximity.
(E) methane concentrations are not related to their proximity to drinking water well location.

## *Free-Response Question*

**Read the following excerpt from a U.S. Environmental Protection Agency press release and answer the questions below.**

> During the late 1980s, an environmental controversy began in Western Oregon, Washington, and Northern California that continues today. At stake were thousands of jobs and the future of old-growth (never been logged) coniferous forests, along with the many species that depend on these forests. Current estimates are that approximately 8% of old-growth forests remained in 2013. The endangered northern spotted owls depended upon preserving the old-growth forests of this region. The federal government proceeded to suspend logging in the 3 million remaining acres where the last 2100 owls live. The timber industry estimates that this halt to logging would cost them 33 billion dollars, with most of the losses borne by local logging companies and residents in Washington and Oregon. Conservationists dispute these numbers of economic loss, stating that there could actually be a *gain* in economic value, also adding that the northern spotted owl is an important umbrella species.

a)  Explain the scientific reasoning and rationale for conservationists claiming an *economic gain* rather than an economic loss with this federal decision.

b)  Describe one specific piece of environmental legislation that could apply in the above conflict, including the rationale of your choice.

c)  Describe one *internal cost* and one *external cost* that might be associated with the timber industry, including your explanation.

d)  Explain what Gross Domestic Product (GDP) is a measure of in the above conflict. Give one criticism of GDP as a progress indicator and provide a better progress indicator, including what it attempts to measure.

# Human Population Dynamics

## *Chapter 8: Human Population*

---

**YOU MUST KNOW**

- The scope of human population growth.
- The divergent views on population growth.
- How to evaluate the effects of human population growth, affluence, and technology on the environment.
- The fundamentals of human demography and the interpretation of age structure diagrams.
- The fundamentals of demographic transition.
- How poverty, the status of women, family planning, and affluence affect population growth.
- How population goals are linked to sustainable development goals.

---

### *The scope of human population growth*

▐ The current world growth rate is approximately 1.2%; roughly 88 million people are added each year.

▐ At a 1.2% growth rate, the world population will double in approximately 58 years (**doubling time = 70 ÷ % growth rate**).

▐ The current world population, as of November 2016, is in excess of 7.3 billion.

### *The divergent views on population growth*

▐ Current world population numbers are due to the many factors that have brought down death rates and infant mortality rates: improved sanitation, better medical care, increased agricultural output, and other technological innovations.

▐ Global birth rates have not declined as much as death rates; therefore, human population has continued to rise.

▪ **Thomas Malthus** (1766–1834), British economist, predicted that human population size would eventually outgrow the available food supply if humans did not purposefully limit births.

▪ **Paul Ehrlich**, in his 1968 book, *The Population Bomb*, predicted population growth would soon lead to famine and that conflict over scarce resources would eventually consume civilization by the end of the 20th century.

▪ These predictions have not fully materialized, primarily because of increased food production innovations, enhanced prosperity, and education/gender equality for women.

▪ Many nations are questioning whether continued population growth is positive, especially if it depletes resources, stresses social systems, and degrades the natural environment.

▪ Population decline and population aging in some countries have given rise to fears of future economic decline. In 2011, the U.S. birth rate had reached its lowest level in history, and two of every three European governments now feel their birth rates are too low to sustain their economic growth.

### *How to evaluate the effects of human population growth, affluence, and technology on the environment*

▪ The $I = P \times A \times T$ model summarizes how environmental impact (I) results from the interactions among population size (P), affluence (A), and technology (T).

▪ Environmental impact is most often equivalent to negative pollution effects and/or resource consumption.

▪ Understanding this model clearly demonstrates that negative environmental effects are not caused solely by the numbers of humans (population size) in a given area but are caused by what those numbers of humans are capable of doing (their rate of consumption of resources due to affluence and technology use).

▪ Scientific studies have shown that humans are responsible for using approximately 24% of the planet's NPP (net primary productivity)—a very large amount for a single species.

▪ Technology advances have allowed for increased efficiency, temporarily alleviating our strain on resources; however, all matter (land, water, natural resources) is finite, and limits to population growth exist.

▪ The size of Earth's carrying capacity is often debated with estimates ranging from a conservative 1–2 billion humans living prosperously in a healthy environment, to 33 billion living in extreme poverty with intensive cultivation without natural areas.

▪ China's growth in total population (currently 1.37 billion) reflects an unprecedented environmental impact (reduced air quality, overdrawn aquifers, high levels of soil erosion) resulting from increased affluence (A) and technology (T). The U.S. total population (currently at 331 million), although significantly lower than China's, has a much higher level of affluence (A) and use of technology (T), resulting in a more significant impact (I) on the earth.

### *The fundamentals of human demography and the interpretation of age structure diagrams*

▌ The principles of population ecology, which apply to animal and plant life, also apply to humans. **Demography** is the study of human populations, how they change with time, and the main factors causing this change.

▌ The following are the main factors useful for predicting population dynamics and environmental impacts.

■ **Population size**, the absolute number of individuals, is only one aspect in determining population dynamics.

■ **Population density** and **distribution** reflect the spatially uneven nature of human population on Earth. Temperate, subtropical, and tropical areas have the highest population density; therefore, these regions often bear the highest environmental impact.

■ **Age structure diagrams (population pyramids)** describe the number of individuals of each age class within a population and are especially valuable for predicting future population change. The width of each horizontal bar represents the number of people in each age class; a wide base indicates a population capable of future rapid growth or population momentum. An age structure diagram with approximately equal age classes from pre-reproductive through post-reproductive ages will remain stable and experience insignificant growth.

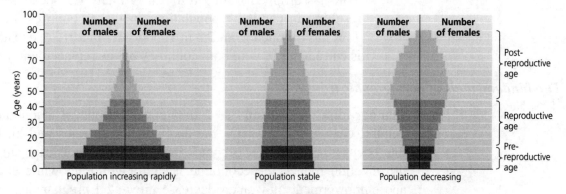

**Figure 8.9** Age Structure Diagrams and Their Ability to Predict Human Population Change

■ **Sex ratios**, the ratios of males to females in a population, affect population dynamics. In most countries, this ratio is approximately equal when people reach reproductive age; however, currently in China, this ratio is strongly skewed in favor of males (because of cultural preferences and China's one-child policy).

- **Birth and death rates**, **immigration**, and **emigration** also determine whether a population grows, shrinks, or remains stable.
  - Birth and immigration *add* individuals to a population.
  - **Crude birth rate (CBR)** is the number of births per 1,000 individuals per year.
  - Death and emigration *remove* individuals from a population.
  - **Crude death rate (CDR)** is the number of deaths per 1,000 individuals per year.
  - *Global* population growth rate $= \text{CBR} - \text{CDR} \div 10$
- A *nation's* growth rate $=$
  $$\frac{(\text{CBR} + \text{immigration}) - (\text{CDR} + \text{emigration})}{10}$$
- or, shown another way:
  $$\frac{(\text{births} - \text{deaths}) + (\text{immigration} - \text{emigration})}{10}$$
- **Total fertility rate (TFR)** is the average number of children born per woman during her lifetime. **Replacement fertility** is the TFR that keeps the size of a population stable $(\text{TFR} = 2.1)$, and if it drops below 2.1, the population size will shrink (in the absence of immigration).
  - Improved medical care (reducing infant mortality), increasing urbanization, and greater educational opportunities for women have recently pushed TFR downward in many nations.
  - TFR has dropped from 2.6 to 1.4 in many European countries, resulting in declining populations in these countries. In China, TFR has dropped from 5.8 in 1950 to 1.7 in 2015. The U.S. TFR is currently 1.9, with both countries below replacement fertility.

## The fundamentals of demographic transition

- **Demographic transition** is a theoretical model of economic and cultural change that explains the declining death rates and birth rates that occur when a nation becomes industrialized. Industrialization and urbanization reduce the economic need for children, while education and empowerment of women decrease unwanted pregnancies, causing growth rates to fall naturally.
- There are four stages of the demographic transition model.
  - **Pre-industrial stage**—death rates are high (disease is widespread), and birth rates are also high (people are compensating for high infant mortality).
  - **Transitional stage**—characterized by declining death rates due to improved medical care and food production.
  - **Industrial stage**—birth rates decline as a result of increased opportunities for women and access to birth control.
  - **Post-industrial stage**—the final stage in which both birth and death rates have fallen to low and stable levels.

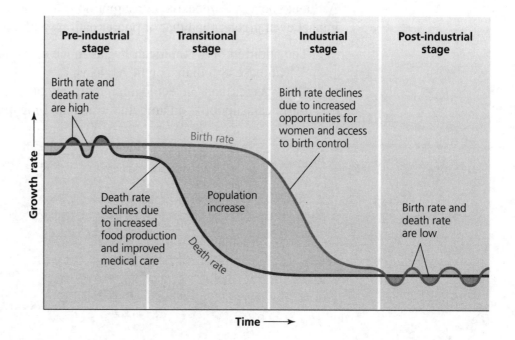

**Figure 8.15**  The Demographic Transition Model and Its Major Stages

▌ Although many European countries, the United States, Canada, and Japan have already passed through demographic transition, it remains an unanswered question for Earth's environment whether developing nations will pass through these stages in the same manner as Western countries.

### How poverty, the status of women, family planning, and affluence affect population growth

▌ Poorer societies tend to show faster population growth than do wealthier societies. Over 99% of the next 1 billion people added to Earth's population will be born into developing countries.

▌ Increasing female literacy is strongly associated with reduced birth rates in many nations.

▌ Perhaps the greatest single factor enabling a society to slow its population growth is the ability of women and couples to practice birth control with contraception.

    ■ 56% of married women worldwide report using a modern method of contraception to plan or prevent pregnancy.

    ■ China has the highest rate of contraception use (84%) of any nation, slowing its growth rate down to 0.5%, primarily through its one-child policy.

    ■ Thailand has been successful in slowing its growth rate to 0.5% by using an education-based approach to family planning and increasing the availability of contraceptives.

    ■ Many African nations have rates below 10% contraception use, where the region's TFR is 4.7 children per woman.

▌ Affluent societies' intensive consumption often makes their ecological impact even greater than that of poorer nations with larger populations.

  ▪ The addition of 1 American has as much environmental impact as 3.4 Chinese, 8 Indians, or 14 Afghans.
  ▪ Our species has an ecological footprint 50% greater than what Earth can support—an amount termed the **biocapacity**.

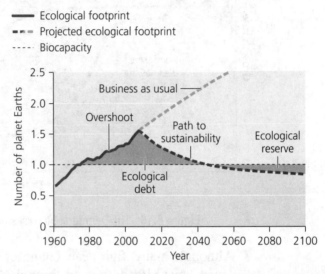

**Figure 8.20** The Global Ecological Footprint of Human Population

▌ Epidemics (such as HIV/AIDS), claiming many young and productive members of society, influence population dynamics and can have severe social and economic ramifications. Africa is being hit the hardest and is struggling to advance through the demographic transition.

▌ China's past restrictive reproductive policies (their one-child policy) have led to a shrinking workforce, a rising number of older citizens, and an unbalance of the sex ratio (many fewer females). Although overall affluence has improved, future productivity of the nation may be compromised.

### *How population goals are linked to sustainable development goals*

▌ In 2000, world leaders came together to adopt the UN **Millennium Development Goals**, which require attention to population growth and resource consumption.

▌ Sustainability demands we stabilize our population size in time to avoid destroying the natural systems supporting our economies and societies.

▌ Currently, there are two major reasons to be encouraged.

  ▪ The rate of population growth has decreased nearly everywhere, even though global population is still rising.
  ▪ Progress has been made in expanding rights for women worldwide, which has significant implications for slowing population growth.

*Multiple-Choice Questions*

1. To determine the number of individuals that will be added to a population in a specified time, one would multiply the growth rate by the
   (A) number of immigrants.
   (B) crude birth rate.
   (C) original population size.
   (D) crude birth rate + immigration.
   (E) final population size.

2. If a country's age structure diagram looks like a pyramid, the population
   (A) will expand for at least one generation.
   (B) is demonstrating exponential growth.
   (C) is shrinking.
   (D) is expanding.
   (E) is stable and remaining the same.

3. Which best describes the reason for the very rapid growth of the human population since the Industrial Revolution?
   (A) Humans have few offspring, but their young survive at a very high rate.
   (B) Environmental resistance does not apply to humans.
   (C) The biotic potential for humans has increased.
   (D) Climatic change favorable to humans has allowed our species to expand its range.
   (E) Technology has allowed the species to overcome some environmental resistance.

4. What is the first significant change that occurs in a country undergoing demographic transition?
   (A) Birth rates decline as a result of improved education for women.
   (B) Birth rates decline and emigration occurs, resulting in a gradually declining population.

   (C) Death rates decline as a result of improved food production and medical care.
   (D) Death rates and birth rates diverge, leading to a gradual population decline.
   (E) Technological advances lead to an immediate population decline and improved status for the general population.

5. If the global population has a CDR of 12 and a CBR of 15, this would result in a growth rate of ____%.
   (A) 0.3
   (B) 3.0
   (C) 30.
   (D) .03
   (E) Cannot determine with the information provided.

6. The *most critical factor* in controlling human population growth is
   (A) decreasing the average number of births per woman.
   (B) controlling the length of one's reproductive lifespan.
   (C) decreasing the age of first birth for a woman.
   (D) decreasing infant mortality.
   (E) increasing medical care and overall wellness for the human population.

7. When the IPAT model is applied to China, the *most* important recent factor that has changed and is instrumental in determining the environmental effects of China's population is
   (A) the value of P.
   (B) the value of A.
   (C) the value of T.
   (D) the value of TFR.
   (E) the value of China's growth rate.

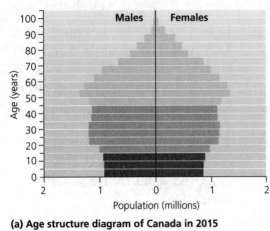

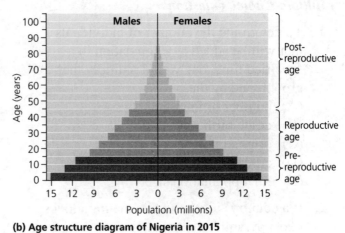

(a) Age structure diagram of Canada in 2015

(b) Age structure diagram of Nigeria in 2015

**Figure 8.10**

*Use the above figure to answer questions 8–10.*

8. Looking at the figure, you are able to determine that
   - (A) the population in (a) is growing rapidly.
   - (B) the population in (b) is declining rapidly.
   - (C) the population in (b) is growing rapidly.
   - (D) the population in (a) is aging rapidly.
   - (E) both countries have completed their demographic transition.

9. What is the population in part (b) likely to experience in the near future?
   - (A) resource depletion and decreased quality of life
   - (B) a high infant mortality
   - (C) lack of economic growth
   - (D) a quick move through demographic transition
   - (E) a decline in total fertility rate

10. The population in part (a) suggests that its total fertility rate (TFR) is
    - (A) less than replacement level.
    - (B) zero.
    - (C) over 4.0.
    - (D) steadily rising.
    - (E) now rising but has been recently declining.

11. When the IPAT model is applied to the United States, the *least* important factor in determining the environmental effects of the U.S. population is
    - (A) the value of P.
    - (B) the value of A.
    - (C) the value of T.
    - (D) the value of A and T.
    - (E) the value of TFR.

12. Replacement fertility
    - (A) restores the population size after a catastrophic event.
    - (B) is a contraceptive technique.
    - (C) is below 2.0 in Latin America and the Caribbean.
    - (D) is below 2.1 in Africa.
    - (E) is equal to 2.1 in stable populations.

13. If a population doubles in the course of 30 years, its growth rate would be close to ____%.
    - (A) 4.6
    - (B) 2.3
    - (C) 5.0
    - (D) 10
    - (E) 1.15

70/GR = 30

GR = 30/70

14. The United States age structure diagram
    (A) looks like a pyramid.
    (B) reflects unequal distribution of males and females at all age groups.
    (C) reflects a "baby boom" in the early 1970s.
    (D) reflects an aging population.
    (E) reflects a population with a high growth rate.

15. The world population growth rate is currently close to ___%.
    (A) 5.5
    (B) 2.1
    (C) 1.2
    (D) 0.5
    (E) 0.25

16. Declining death rates due to increased food production and improved medical care, with birth rates that remain high, is characteristic of the _____ stage.
    (A) pre-industrial
    (B) stabilization
    (C) transitional
    (D) post-industrial
    (E) revolutionary

*Free-Response Question*

In January of 2015, the population of China was estimated to be approximately 1.37 billion, with an annual growth rate of 0.5%. In 1949, when Mao Zedong founded the country's current political regime, approximately 500 million people lived in China, the total fertility rate (TFR) was 5.8 children, and the country was mostly rural, undeveloped, and impoverished. Include all calculations and provide full explanations for the following questions.

a) Calculate the *% change* in China's population from 1949 to 2015.

b) Describe/explain the *stages* of demographic transition China has experienced. Include your specific evidence for each stage.

c) Describe the approximate shape of the base (pre-reproductive age class) of China's age structure diagram in 2015, as compared to the size of the reproductive and post-reproductive age classes. Include your evidence.

d) Calculate the number of *additional* people China will have at the end of 2015.

e) Based on the current growth rate, calculate the *doubling time* for China's population.

# Earth's Land Resources and Use

## Chapter 9: Soil and Agriculture

---

### YOU MUST KNOW

- The importance of soils to agriculture.
- The impacts of agriculture on soils.
- The major developments in the history of agriculture.
- The fundamentals of soil science, including soil formation and soil properties.
- The major causes of soil erosion and land degradation.
- The principles of soil conservation and specific solutions to modern agriculture's overuse of water and fertilizer.
- The importance of pollinators to crop success.
- The major policy approaches for pursuing soil conservation and sustainable agriculture.

---

*The importance of soils to agriculture*

- Healthy soil is vital for agriculture, for forests, and for the functioning of Earth's natural ecosystems, and is especially important as the foundation for sustainable agriculture.
- **Soil** is a complex system of inorganic, weathered parent material (rock), organic matter, water, gases, nutrients, and microorganisms. Soil is a renewable resource; however, if soil is degraded or depleted, this renewal process occurs very slowly.
- More than one out of every three acres of land on Earth produces food and fiber (cropland covers 12% of Earth's land surface and rangeland covers 26%).

## The impacts of agriculture on soils

▌ Each year, Earth gains more than 80 million people, yet loses 12 to 17 million acres of productive **cropland** through a process termed land degradation.

▌ **Soil degradation** has resulted primarily from forest removal, cropland agricultural practices, and overgrazing of livestock. In the United States, approximately 5 tons of topsoil are lost for every ton of grain harvested.

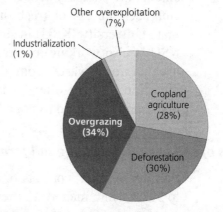

**Figure 9.13a**   Causes of Soil Degradation

## The major developments in the history of agriculture

▌ Approximately 10,000 years ago, as glaciers retreated and the climate warmed, people in some cultures began to raise plants from seed, to domesticate animals, and to practice selective breeding (choosing the best/hardiest versions of wild plants) by intentionally planting the most desirable seeds.

▌ The Middle East's "Fertile Crescent," China, Africa, and the Americas each began traditional agricultural practices independently. Initially, these farming practices were **subsistence agriculture**, producing only enough food for the family farming the land.

▌ The Industrial Revolution introduced large-scale mechanization and fossil fuel combustion to agriculture in the 19th century, boosting crop yields and also intensifying irrigation demands. Synthetic fertilizers and pesticides were introduced in the 1940s, along with the practice of **monoculture** (large-scale, uniform planting of a single crop). This approach was quite distinct from the **polyculture** ("many types") practices of traditional agriculture that mixed maize, beans, and squash in the same fields.

▌ Beginning in the 1950s, the **Green Revolution** spread industrial agriculture techniques as practiced in developed nations to developing countries. These new mechanized technologies and seed varieties increased crop yields, helping to avoid mass starvation in many developing countries; however, much soil degradation has resulted from the intensive use of synthetic fertilizers, pesticides, and the monoculture approaches of the Green Revolution.

▌ Today, **industrial agriculture** dominates Earth's land use by humans and has many negative consequences, including degradation of soil, water, and pollinators that we rely on for terrestrial food supply.

▌ **Sustainable agriculture** has characteristics that maintain healthy soil, clean water, pollinators, and other resources that are necessary for *long-term* crop and livestock production.

▌ University and college dining services around the country are now among the industry leaders in **culinary sustainability**, a pursuit that embraces the use of fresh, healthy, locally produced foods. Kennesaw State University (KSU) in suburban Atlanta, GA, is a leader in this area.

▌ KSU is also working to create a fully **closed loop system** in which the uneaten food waste from dining commons is fed into an anaerobic digestion system, where it is broken down to generate a nutrient-rich liquid and then used to nourish soils of the campus cropland.

### *The fundamentals of soil science, including soil formation and soil properties*

▌ **Soil**, by volume, consists of approximately 50% mineral matter and up to 5% organic matter. The remainder consists of pore space taken up by air or water. The organic matter includes living and dead microorganisms as well as decaying material derived from plants and animals.

▌ Because of the complex interactions between its living and nonliving components, soil meets the definition of an ecosystem.

▌ Soil formation is the slow physical, chemical, and biological **weathering** of **parent material** (the base geologic material, rock, or sediment of a particular location) and the accumulation and transformation of organic matter.

▌ The main factors influencing soil formation are climate, parent material composition, organisms, topography, and time.

▌ **Soil horizons**, or distinct layers of soil, eventually develop as wind, water, and organisms move and sort the fine particles weathering creates, producing a soil profile. The simplest way to categorize horizons is to recognize the A (topsoil), B (subsoil), and C (parent material) horizons. However, soil scientists often add at least three additional horizons (O, E, and R).

  ▪ **O horizon**—the litter layer, consisting mostly of organic matter.
  ▪ **A horizon**—the topsoil, consisting of some organic material mixed with mineral components.
  ▪ **E horizon**—the zone of leaching (eluviation), where minerals and organic matter tend to leave this horizon and move into the next lower layer.
  ▪ **B horizon**—the subsoil, where minerals and organic matter accumulate from above.
  ▪ **C horizon**—consists largely of weathered parent material.
  ▪ **R horizon**—the rock or sediment of pure parent material.

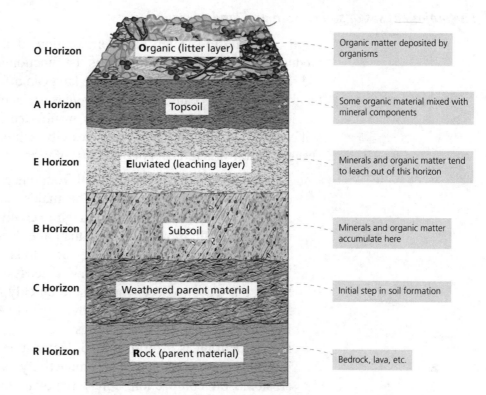

| O Horizon | Organic (litter layer) | Organic matter deposited by organisms |
| A Horizon | Topsoil | Some organic material mixed with mineral components |
| E Horizon | Eluviated (leaching layer) | Minerals and organic matter tend to leach out of this horizon |
| B Horizon | Subsoil | Minerals and organic matter accumulate here |
| C Horizon | Weathered parent material | Initial step in soil formation |
| R Horizon | Rock (parent material) | Bedrock, lava, etc. |

**Figure 9.4**  The Major Soil Horizons

- The A horizon, or **topsoil**, is the crucial horizon for agriculture and ecosystems. It is the most nutritive, taking its texture, coloration, and water-holding capacity from the **humus** (partially decomposed organic matter) content.

- Soil scientists classify soils based largely on properties of color, texture, structure, pH, and cation exchange.

  - **Soil texture** is determined by the size of its particles, ranging from **clay** (the smallest) to **silt** (intermediate), to **sand** (the largest). **Loam** is soil with an even mixture of clay, silt, and sand.

  - **Cation exchange capacity (CEC)** is the process by which plants' roots donate hydrogen ions to the soil in exchange for cations (positively charged ions) of nutritional value such as calcium, magnesium, and potassium. CEC is often referred to as the nutrient-holding capacity of soil.

- Soils with fine texture (clay) and also rich in organic matter have a high CEC. However, if a soil has more than 20% clay, its water retention becomes too great, the soil becomes waterlogged, and plant roots are deprived of oxygen. Also, acid soils with a low pH have a low CEC, allowing essential nutrients to leach away, such that they are unable to reach the plant to promote healthy growth.

### The major causes of soil erosion and land degradation

- **Land degradation** refers to deterioration that diminishes land's productivity and biodiversity, impairs the functioning of its ecosystems, and reduces the ecosystem services the land can offer.
- Land degradation is caused by intensive/unsustainable agriculture, deforestation, and urban development. It is manifested in processes such as soil erosion, nutrient depletion, water scarcity, salinization, waterlogging, changes in soil structure and pH, and loss of organic matter from the soil.
  - **Erosion** is the removal of material from one place and its transport to another by wind or water. The major causes of soil erosion are: 1) overcultivating fields through excessive tilling; 2) grazing rangeland with more livestock than the land can support; and 3) removing forests on steep slopes or with large clear-cuts.
  - Soil erosion is a major problem for ecosystems and agriculture because it tends to occur much more quickly than soil is naturally formed, and it also tends to remove topsoil (A horizon), the most valuable soil layer for living things.
  - **Desertification** is a form of land degradation affecting arid lands in which more than 10% of productivity is lost as a result of erosion, soil compaction, forest removal, overgrazing, drought, salinization, water depletion, or other factors. China and Northern Africa are experiencing increasing levels of desertification and gigantic dust storms.
  - In the early 1930s, a major drought occurred along with the already large-scale cultivation and overgrazing of native grasses in North America's Great Plains. The resulting **Dust Bowl** forced thousands of farmers off their unproductive land as millions of tons of topsoil eroded and were lost forever.
  - **Overgrazing** by cattle, sheep, goats, or other livestock can contribute to soil degradation, especially when native grasses are removed. Worldwide, overgrazing causes as much soil degradation as conventional cropland agriculture does, and it is an *even greater* cause of desertification.
  - To restore grasslands, scientists have now collected sufficient evidence from the McKinney Flats Project (located in southern AZ and NM, the largest replicated terrestrial ecological experiment in North America) indicating the positive role of controlled burns (fire).

### The principles of soil conservation and specific solutions to modern agriculture's overuse of water and fertilizer

- **The Soil Conservation Act** (1935) established the Soil Conservation Service (SCS) in response to the devastation caused in the Dust Bowl. In 1994 it was renamed the **Natural Resources Conservation Service (NRCS)** and its responsibilities were expanded to include water quality protection and pollution control.

▌ Soil conservation efforts are thriving internationally, often utilizing a variety of approaches. Farming methods making use of the general principle of *maximizing vegetative cover to protect soils against erosion* are increasing, as are more sound methods for the use of water.

- **Conservation tillage** describes an array of approaches that reduce the amount of tilling relative to conventional farming. It is any method of limited tilling that leaves more than 30% of crop residue covering the soil after harvest. Critics have noted that this approach often requires substantial chemical herbicide and synthetic fertilizer use because weeds are not physically removed from fields after crop harvest.

- **No-till agriculture** is the most extreme form of conservation tillage and does not involve tilling (plowing or disking) the soil. A "no-till drill" cuts furrows through the O horizon of dead weeds and crop residue and the upper levels of the A horizon, dropping seeds and fertilizer into the furrow. By not removing the previous year's crop residue, the increased organic matter also helps to improve soil's fertility.

  - Researchers today are debating just how much impact no-till farming can have on carbon sequestration and global climate change mitigation; however, studies to date seem to have determined that natural ecosystems sequester more carbon than croplands and that no-till farming is not a *major* solution to climate change, although it does help store carbon.

  - Nearly 25% of American farmland is under no-till cultivation and over 40% of American farmland is using some form of conservation tillage. Although pioneered in the United States and the United Kingdom, no-till and conservation tillage methods are now most widespread in subtropical and temperate South America (Brazil, Argentina, and Paraguay).

  - No-till critics in the United States note that this approach often requires heavy use of chemical herbicides (because weeds are not physically removed) and synthetic fertilizers (non-crop plants can take up significant soil nutrients).

- **Crop rotation** methods involve alternating the type of crop grown in a given field from one season to the next. This returns nutrients to the soil (legume planting), breaks plant disease cycles, and minimizes erosion.

- **Contour farming** consists of plowing furrows sideways across a hillside, perpendicular to its slope and following the natural contours of the land to prevent runoff and erosion.

- **Terracing** is the most effective method for preventing soil erosion on extremely steep terrain. Terraces are level platforms, or steps, that are cut into steep hillsides to contain irrigation water.

- **Intercropping** minimizes erosion by planting different types of crops in alternating bands or other spatially mixed arrangements. The additional ground cover slows erosion much more readily than a single crop.
    - **Shelterbelts**, or windbreaks, are a widespread technique used to reduce erosion caused by wind. These consist of rows of trees or tall shrubs that are planted along the edges of fields to slow the wind.
- While water is essential for maximum agricultural productivity, **irrigation**, the artificial provision of water beyond that which crops receive from rainfall, can also damage soil (70% of all fresh water used on Earth is for irrigation). Irrigation efficiency worldwide is low, with plants using less than half of all the water that is applied.

    - **Salinization** occurs most frequently in arid agricultural areas when the small amounts of salts in irrigation water become highly concentrated on the soil surface through high evaporation rates. Salinization now inhibits production on one-fifth of all irrigated cropland globally, costing more than $11 billion each year. Irrigating efficiently, planting crops requiring less water, and irrigating with water low in salt help to prevent salinization.
    - **Waterlogging**, another water-use issue, occurs when over irrigation saturates the soil and causes the water table to rise to the level that plant roots are drowned, essentially suffocating the plants.
    - When considering water-use solutions, sustainable approaches better match crops to climate (e.g., not growing rice and cotton in deserts) and use **drip-irrigation** systems that target water directly toward plants (hoses drip water directly into soil near the plant's roots).

- Nutrients for plants are also essential, along with adequate water and a healthy soil. However, **fertilizers**, especially when over-applied, can also chemically damage soils. Inorganic (mined or synthetically manufactured mineral supplements of N, P, or K) fertilizers are more susceptible to leaching and runoff, causing significant effects in waterways (eutrophication leading to hypoxia and "dead zones"). Organic (remains or wastes of organisms, manure, crop residues, and compost) fertilizers can improve soil structure, nutrient retention, and water-holding capacity, helping to prevent erosion.
- Sustainable approaches to fertilizing crops target the delivery of nutrients to plant roots and avoid the over application of fertilizer. Planting buffer strips of vegetation along field edges and along rivers/streams on farmland property help to capture nutrient runoff along with any eroded soil.

### *The importance of pollinators to crop success*

- Although our grain crops (wheat and corn) are wind-pollinated, many fruit, vegetable, and nut crops depend on insects for pollination. In contributing to food production success, these insects provide more than $150 billion in ecosystem services.
- At least 800 types of cultivated plants rely on bees and other insects for pollination, with an estimated 73% relying directly on honeybees.
- The European honeybee, *Apis mellifera*, forms the foundation of modern agriculture, pollinating more than 100 crops that make up one-third of the U.S. diet.
- As a result of increased pesticide use and habitat loss, populations of native pollinators are decreasing and even disappearing from many regions. Finding ways to conserve pollinators is an essential step to attaining sustainable agriculture.

### *The major policy approaches for pursuing soil conservation and sustainable agriculture*

- Today's industrial agriculture exerts tremendous impacts (often negative) on Earth's terrestrial lands and ecosystems. Some existing government policies often *worsen* land degradation.
  - **Subsidies** (government incentives of cash or tax breaks) can encourage people to cultivate land that would otherwise not be farmed. Approximately one-fifth of the income of the average U.S. farmer comes from subsidies.
  - American ranchers currently benefit from subsidies, paying substantially **low grazing fees** to graze livestock on Bureau of Land Management (BLM) lands. In 2016, a grazing permit was just $2.11 per month per "animal unit." Such low fees can encourage overgrazing and land degradation through a "tragedy of the commons" scenario.
  - **Wetlands** have been drained for agricultural purposes, resulting in a loss of over 50% of all the original wetlands in North America.
- The **Conservation Reserve Program**, established in the 1985 farm bill, pays farmers to stop cultivating highly erodible cropland and instead places it in conservation reserves planted with grasses and trees. Since 2007, the area in conservation reserves has decreased by 35% in response to higher food prices, as farmers have withdrawn lands from the program and planted with crops or grazing grasses.
- The *Wetlands Reserve Program* provides a new view of the valuable ecosystem services of wetlands in the U.S. Financial incentives are being used as a means to protect wetlands, offering payments to landowners who restore or enhance wetland areas on their property. Over 2 million acres are currently enrolled, and this program is funded by the U.S. Congress in the latest farm bill legislation.

Internationally, the United Nations promotes soil conservation and sustainable agriculture through a variety of programs led by the Food and Agriculture Organization (FAO). Rather than following a top-down government-mandated approach, the Farmer-Centered Agricultural Resource Management Program (FARM), operating in eight Asian countries, calls on the creativity of local communities to educate and encourage farmers to conserve soils and secure their food supply.

# Chapter 10: Agriculture, Biotechnology, and the Future of Food

---

## YOU MUST KNOW

- The challenge of feeding a growing human population.
- The goals, methods, and consequences of the Green Revolution.
- The environmental issues involved in raising animals for food.
- The approaches for preserving crop diversity.
- The threats to pollinators with potential solutions.
- The strategies for pest management.
- The characteristics and potential of organic agriculture.
- The science behind genetic engineering and debate over genetically modified food.
- The environmental imperative of sustainable agriculture.

### The challenge of feeding a growing human population

- Global food production has risen more quickly than world population growth over the past half-century.
- This steady rise in food production has resulted from the increased use of fossil fuels; a highly intensified use of irrigation, pesticides, and fertilizers; the cultivation of more land; and the development (through selective breeding and genetic engineering) of more productive crop and livestock varieties.
- Despite rising food production, nearly 800 million people experience **undernutrition** (receiving fewer calories than the minimum dietary energy requirement); over 1 billion people are overweight as a result of **overnutrition**, and millions of children experience **malnutrition** (a shortage of essential nutrients, such as protein).

### The goals, methods, and consequences of the Green Revolution

- In the 1940s, the desire for a greater quantity and quality of food for the world's growing population led **Norman Borlaug** (who later won the Nobel Peace Prize for this work) to develop improved varieties of wheat through selective breeding. This technology was quickly transferred to developing countries such as Mexico, India, and Pakistan.

- Along with the new strains of wheat, rice, and corn, the Green Revolution methods required synthetic fertilizer and pesticides and they increased water and fossil fuel consumption, often resulting in localized pollution, soil erosion, salinization, and desertification.

- **Monocultures**, large expanses of single crop types replaced the traditional, small-scale polycultures as they were found to be more efficient for planting and harvesting; however, the risk of catastrophic failure of an entire crop due to disease increased, especially without large applications of pesticides.

- Between 1960 and 2008, food production rose 150% and population rose 100%, while land area converted for agriculture increased only 10%.

- As industrial agriculture practices and the planting of monocultures increased worldwide, **seed banks** (institutions preserving genetic diversity as a kind of living museum) were initiated, storing millions of seeds from around the world as a safeguard against global agricultural collapse.

- **The Sierra de Manantlán Biosphere Reserve** (southern Mexico) is also an important repository of crop biodiversity, especially for maize (corn). Preserving such traditional varieties of crops in their ancestral homelands is important for securing the future of the world food supply, as these varieties serve as essential reservoirs of genetic diversity, which we may need to sustain or advance future agriculture.

### The environmental issues involved in raising animals for food

- Global meat production has increased fivefold since 1950, and per capita meat consumption has doubled.

- Food choices have significant implications for energy, land, and water usage; therefore, what we eat can significantly affect environmental quality.

- Different animal food products (beef, pork, chickens, eggs, and milk) require different amounts of energy input (animal feed). The energy input per calorie of food produced is called the **energy subsidy**. For example, it takes 20 kg of grain to feed to cattle to produce 1 kg of beef (an energy subsidy of 20); whereas, it takes only 2.8 kg of grain to produce 1 kg of chicken (an energy subsidy of 2.8).

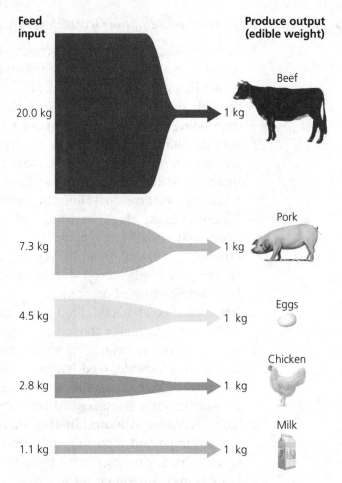

**Figure 10.8** Animal Feed Inputs for Various Produce Outputs

▌ **Feedlots**, an industrial agriculture practice also known as factory farms or **concentrated animal feeding operations(CAFOs)**, are huge warehouses or pens designed to deliver energy-rich food to confined animals living at extremely high densities. Farmers minimize land costs, improve feeding efficiency, and increase the fraction of food energy that goes into production of animal body mass. Heavy use of antibiotics (to control disease) and hormones (given to encourage fast weight gain) are standard feedlot practices.

▌ One-third of the world's cropland is devoted to growing feed for animals, and 45% of our global grain production goes to livestock and poultry.

▌ High-density animal farming leads to a number of environmental and health effects. The FAO recently estimated that U.S. livestock accounted for 55% of the nation's soil erosion, 37% of pesticide application, 50% of the antibiotics consumed, and one-third of the phosphorus pollution of U.S. waterways.

- Surface water and groundwater become polluted from animal wastes (feedlot runoff), resulting in eutrophication of localized waters.
- Many disease-causing microorganisms such as *Salmonella* and *E. coli* can be released in animal wastes and end up in water supplies.
- The heavy use of antibiotics and growth hormones, much of which are excreted by the animals, end up in wastewater and may be transferred up the food chain to humans, causing health problems including increased antibiotic resistance of microorganisms and human cancers.
- Livestock are a major source of greenhouse gases that lead to climate change (livestock account for 5% of $CO_2$ emissions, 44% of our methane emissions, and 53% of our nitrous oxide emissions). Overall, 14.5% of the emissions driving climate change—a larger share than automobile transportation!

- Wild fish populations have declined precipitously worldwide as increased demand (due to human population growth) and new fishing technologies have led to overharvesting. **Aquaculture**, the cultivation of aquatic organisms for food in controlled environments, is now the fastest-growing type of food production: Global output has doubled in just the last decade.

- The environmental and human health *benefits* of aquaculture include the following.

  - Ensures a reliable protein source for human consumption
  - Helps to reduce fishing pressure on overharvested and declining wild fish stocks, as well as lessens the **bycatch** (unintended catch of nontarget species)
  - Uses much less fossil fuel than does commercial fishing

- The environmental *disadvantages* of aquaculture include the following.

  - Aquaculture wastewater may contain bacteria, viruses, and pests that thrive in these high-density habitats and can then infect wild fish and shellfish populations.
  - Aquaculture wastewater contains much organic matter that can lead to eutrophication of local waters.
  - Fish that escape from aquaculture facilities may harm wild fish populations by outcompeting native species or by interbreeding with them (a significant concern if the escapees are GM farmed fish).

## *The approaches for preserving crop diversity*

- Modern industrial monocultures of genetically similar plants have made our global agricultural systems vulnerable to diseases and growing pesticide resistance.
- As industrial agriculture practices have increased worldwide, **seed banks** (institutions preserving genetic diversity as a kind of living museum) were initiated. The **Svalbard Global Seed Vault** (Spitsbergen, Norway) is the most renowned; however, The Royal Botanic Garden's Millennium Seed Bank in Britain, the U.S. National Seed Storage at Colorado State University and the Wheat and Maize Improvement Center in Mexico are also major storehouses of seed diversity.

## *The threats to pollinators with potential solutions*

- **Pollination**, the process by which male sex cells of a plant (pollen) fertilize the female egg cells of a plant, is essential for plant reproduction. Although grain crops are wind-pollinated, many other crops depend on insects including bees, flies, wasps, beetles, moths, butterflies, bats, and birds, producing $15 billion in ecosystem services and invaluable food crops.
- In recent years, up to one-third of all honeybees in the U.S. have vanished from what is being called *colony collapse disorder*. Leading hypotheses as to causes are insecticides, weed-killers (like Roundup that eliminate weeds from farm monocultures), parasitic mites, or a combination of these stresses that weaken bees' immune systems.
- Integrated pest management, biocontrol, planting buffer strips along highways to protect against incursion of insecticides or GM material from conventional farming are all positive actions that can be taken on behalf of pollinators.

## *The strategies for pest management*

- **Pesticides** are synthetic chemicals used to kill insects (insecticides), plants (herbicides), and fungi (fungicides), which humans consider damaging to valuable crops.
- Since 1960, pesticide use has risen fourfold worldwide. This widespread use of pesticides often kills nontarget organisms in addition to the targeted species.
- **Neonicotinoids**, a new class of chemical insecticides, have become common in treating crop seeds. The neonicotinoids become systemic in the plant, ending up in leaves, stems, fruits, and pollen. These "nionics" ultimately make the plant toxic to insects, killing them as they feed on or pollinate the plant.

▌ Pests can evolve resistance to chemical pesticides, requiring humans to design ever-increasingly toxic poisons in order to kill the genetically resistant individuals, a phenomenon known as the "**pesticide treadmill.**"

▌ **Biological control** (biocontrol) is the use of living species (rather than pesticides) that are the natural enemy of or prey upon or parasitize the targeted pest.

■ Parasitoid wasps are frequently used as biocontrol agents of many caterpillars.

■ *Bacillus thuringiensis* (**Bt**) is a naturally occurring soil bacterium that produces a toxin that kills many caterpillars and some fly and beetle larvae. Bt can be sprayed on crops to protect against these insect attacks.

■ Occasionally the biocontrol agent itself can become a pest or have unintended effects if used in an area where it is not native and no natural predators exist. For example, the cactus moth's introduction to the Caribbean islands and its subsequent spread to Florida is of concern to Mexico and the American Southwest, where it could decimate native prickly pear cactus.

▌ **Integrated pest management** (**IPM**) incorporates numerous techniques, including biocontrol, use of limited chemicals when essential, crop rotation, transgenic crops, alternative tillage methods, and mechanical pest removal.

## *The characteristics and potential of organic agriculture*

▌ **Organic agriculture** uses no synthetic fertilizers, insecticides, fungicides or herbicides, but instead relies on biological approaches such as biocontrol and composting.

▌ Worldwide sales of organic food grew 4.5 times between 2000 and 2014, when sales neared $80 billion, as consumers are favoring organic food out of concern about health risks posed by the pesticides, hormones, and antibiotics used in conventional agriculture.

▌ Organic agriculture takes up less than 1% of agricultural land worldwide. This area is rapidly expanding, with Mexico the leader among developing nations. Europe has 28.7 million acres under organic management and the U.S. has 5.4 million acres.

▌ On average, crop yields on organic farms are about 80% of the yield on conventional industrial farms. Greenhouse gas emissions and energy input are noticeably less with organic farming methods than with conventional farming.

▌ The world's two longest-running field experiments in organic farming (in Switzerland and at Rodale Institute, Pennsylvania) have produced soils with more microbial life, earthworm activity, water-holding capacity, topsoil depth, and naturally occurring nutrients than found with conventional farming.

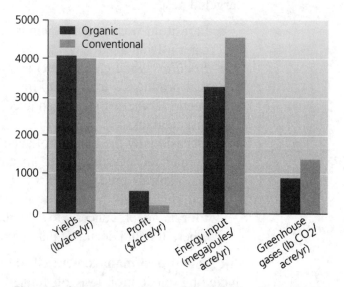

**Figure 2** Organic versus Conventional Farming Comparisons

### *The science behind genetic engineering and debate over genetically modified food*

▌ **Genetic engineering** is any process whereby scientists directly manipulate an organism's genetic material in the laboratory by adding, deleting, or changing segments of its DNA.

▌ **Genetically modified organisms (GMOs)** are organisms that have been genetically engineered using **recombinant DNA**, which is DNA that has been spliced together from multiple organisms. The goal is to place genes that code for certain desirable traits (such as rapid growth, disease resistance, or high nutritional content) into the genomes of organisms lacking those traits. Roundup Ready soybeans are currently the most abundant GM crop in the world.

▌ **Transgenic organisms** are those that contain DNA from another species. All GM organisms are also transgenic species.

▌ The techniques geneticists use to create GM organisms differ from traditional selective breeding in several ways.

- Selective breeding mixes genes from individuals of *same* species; whereas recombinant DNA techniques routinely mix genes of *different* species.

  - Selective breeding deals with *whole* organisms in nature, whereas genetic engineering works with *genetic material* in the lab.
  - Selective/traditional breeding selects from genes that come together on their own, whereas genetic engineering creates combinations that did not previously exist in nature and otherwise would not come together on their own.

- Today, most GM crops are engineered to resist herbicides and/or to resist insect attack. Examples include "**Roundup Ready crops**" (patented by Monsanto Chemical containing genes to resist their widely used herbicide Roundup), **Bt crops** (which have the ability to produce their own toxic effects on insects), and Golden rice (a nutritionally enhanced rice with GM genes producing vitamin A).

- In the United States, 90% of corn, soybeans, cotton, and canola consist of GM strains, and close to half of these crops are engineered for more than one trait. It is conservatively estimated that over 70% of processed foods in U.S. stores contain GM ingredients.

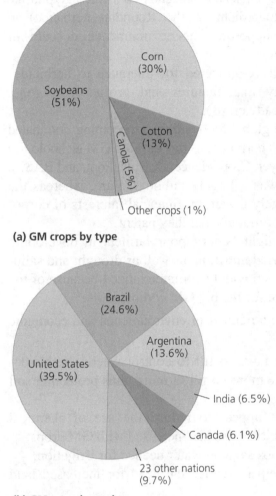

(a) GM crops by type

(b) GM crops by nation

**Figure 10.24 a and b**  GM Crops by Type and GM Crops by Nation

- **Critics of GM crops** cite environmental, health, and ethical concerns.

  - Currently, most scientists feel that *ecological impacts* of GM foods are a greater issue than potential health impacts. It is also known that much of the research into GM food safety is funded and/or conducted by the same corporations that stand to profit from transgenic crop approval.
  - Transgenes from GM crops might escape into nature, harming non-target organisms. There is widespread concern today that the growing organic food market will be hindered if organic farms are contaminated with pollen or seed from GM plants.
  - Pests might evolve resistance to the GM crops and become "super pests."

- Herbicide-tolerant crops tend to result in *more herbicide use* as farmers may apply additional herbicide if their crops can withstand it. Weeds may even evolve resistance to herbicides, making heavier applications necessary. Scientists have recently confirmed glyphosate resistance (the active ingredient in the Roundup herbicide) in 32 species of weeds, supporting the environmental concern of "superweed" proliferation.
- GM crops are genetically engineered for tolerance to herbicides that the same company manufactures and profits from (e.g., Monsanto's Roundup Ready crops).
- There is concern that the global food supply is becoming dominated by a few large agrobiotech corporations that develop GM technologies (Monsanto, Syngenta, Bayer, CropScience, Dow, DuPont, and BASF).
- The Green Revolution was a largely public venture, whereas the "gene revolution" is largely driven by financial interests of corporations selling these GM products that they patent.
- Crops with traits that might benefit poor farmers in developing countries (such as increased nutrition, as well as drought and salinity tolerance) have not been widely commercialized because of the lower economic incentive for the big GM companies.

- **Supporters of GM crops** cite a number of environmental and economic benefits.

  - Because farmers do not have to till to control weeds, herbicide-tolerant crops may enable more no-till farming, thus protecting soil from erosion.
  - Insect-resistant Bt crops appear to reduce the use of chemical pesticides; however, herbicide use has increased with GM crops.
  - Drought-tolerant GM crops save on water needed for irrigation.
  - High-yield GM crops help to reduce the need for increased land clearing.
  - A 2014 meta-analysis of GM crop production found that GM crops produce yields 22% higher than non-GM crops.

- However, many experts feel that with regard to GM foods, we should be employing the **precautionary principle**: One should not undertake a new action until the ramifications of that action are fully understood.
- More than 60 nations require labeling of GM foods, although the U.S. does not. Only a few states (CT, VT, ME) have currently passed GMO labeling laws.

### The environmental imperative of sustainable agriculture

- Industrial agriculture has allowed food production to keep pace with our growing population. However, it has also led to many adverse environmental impacts such as soil degradation; heavy reliance on fossil fuels; and human health concerns over heavy pesticide use, genetic modification, and intensive feedlot operations.

▎**Sustainable agriculture** is agriculture that does not deplete soils faster than they form and does not reduce the amount of clean water or genetic diversity essential to long-term crop and livestock production. Fossil fuel inputs are greatly reduced, as are pesticides, synthetic fertilizers, growth hormones, and antibiotics. Reliance on biological approaches mimicking natural ecosystem functioning are emphasized, including composting, biocontrol, and recycling of matter.

▎**Organic agriculture** is the production of crops without the use of synthetic pesticides or fertilizers, using ecological principles and attempting to keep as much organic matter in the soil as possible. The crops cannot be genetically engineered.

▎The **Organic Food Production Act (1990)** established the national standards for organic products (both crops and livestock) in the United States.

▎Currently, organic agriculture takes up less than 1% of agricultural land worldwide, but it is growing rapidly. Mexico is a leader in organic agriculture among developing nations: Approximately 3% of Mexico's land is used for organic agriculture and 30% of Mexico's coffee crop is now organic.

▎**Locally grown agriculture** is another key component of the move toward sustainable agriculture, reducing fossil fuel use by avoiding long-distance transport. **Community-supported agriculture (CSA)**, where consumers partner with local farmers to pay in advance for a share of their yield, and farmers' markets are also growing rapidly in most communities.

# Chapter 12: Forests, Forest Management, and Protected Areas

**YOU MUST KNOW**

- The main ecological and economic contributions of forests.
- The history and current scale of global deforestation.
- The approaches to sustainable forest management in relation to fire, pests, and current global climate change.
- The major federal land management agencies and the lands they manage and protect.
- The various types of protected land designation, including evaluation of issues involved in their design.

### *The main ecological and economic contributions of forests*

▌ Forests currently cover 31% of Earth's land surface.

▌ The major forest types are boreal forests (primarily conifers of Canada, Scandinavia, and Russia), tropical rainforests, temperate deciduous forests, temperate rainforests, and tropical dry forests.

▌ Because forests are ecologically complex, they provide many niches for organisms; forests are thus some of the richest ecosystems for biodiversity. As forests change over time through succession, the species composition of trees also changes, with old-growth forests containing more biodiversity than younger forests.

▌ Forests supply us with many vital **ecosystem services** such as soil stabilization, flood prevention (through the slowing of rainfall's runoff), recharge of aquifers, water purification of streams and rivers (by acting as pollutant filters), moderation of precipitation patterns and climate (through photosynthesis and transpiration processes), enriched topsoil (by returning organic matter in the form of litter to soil), and carbon storage (in the form of organic matter produced through photosynthesis).

▌ **Carbon storage** is now considered one of the most important ecosystem services because of its connection to global climate change. Carbon dioxide (taken in by all plants during photosynthesis and chemically changed into organic compounds) is the primary greenhouse gas contributing to global climate change. The more forests we preserve or restore, the more carbon (in the form of carbon dioxide) that can be kept out of the atmosphere.

▌ Forests provide many economically valuable resources such as medicines, dyes, fibers, food, and wood. Today, approximately 30% of all forests are designated for timber production.

### *The history and current scale of global deforestation*

▌ **Deforestation**, the clearing and loss of forests, has been driven by the demand for wood (as fuel and building material) and for agricultural land to feed the world's growing population. Currently, forest loss is greatest in the tropical forests of Central America and Africa.

▌ Deforestation fueled the growth of the U.S. westward. Today there is very little **primary forest**, or **old-growth forest** (natural forests uncut by humans or natural disasters in 200 years or more), remaining in the lower 48 U.S. states. **Secondary-growth forests** (forests that have grown back from secondary succession once land has been cleared or cut) have smaller trees and lower species composition than original forests.

▌ Developing nations, in need of economic revenue, are granting **concessions** (the right to extract a resource) to multinational corporations, which clear-cut forests to sell the wood and wood products. For example, throughout Southeast Asia and Indonesia today, primary forest clearing has been used to establish tree plantations of oil palms (palm oil is used in snack foods, soaps, and cosmetics, and also as a biofuel).

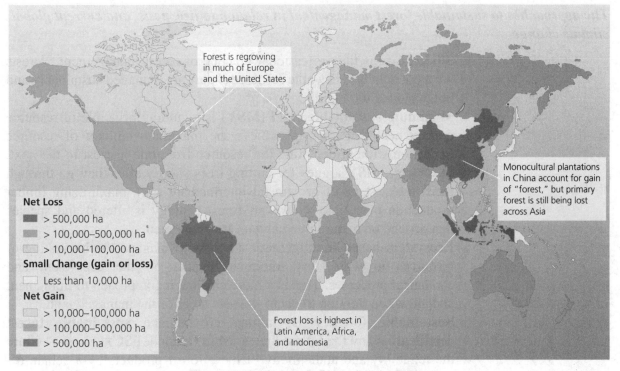

**Figure 12.6b**  Change in Forest Area by Region

Solutions to deforestation are emerging slowly.

- **Conservation concessions** are a type of concession in which a conservation organization buys a concession and uses it to *preserve* the forest rather than cut it down.
- **Carbon offsets** curb deforestation and climate change together. Wealthy industrialized nations would pay poorer developing nations to conserve forests. In the international cap-and-trade system, the developing country would gain income while the industrialized nation would receive carbon credits to offset its emissions.
- **Debt-for-nature swap** is an approach in which a conservation organization raises money and offers to pay off a portion of a developing nation's debt in exchange for that nation setting aside reserves, funding environmental education, and improving management of protected areas. The U.S. participates in this program through its 1998 Tropical Forest Conservation Act, forgiving Indonesia $30 million of debt in return for protecting the forested areas that are home to the Sumatran tiger and other species.

***The approaches to sustainable forest management in relation to fire, pests, and current global climate change***

- **Silviculture** is the professional practice of forest management. Forests are a potentially renewable resource *if* they are managed sustainably and not exploited too rapidly.

- **Maximum sustainable yield (MSY)** is a guiding principle of resource management that aims to achieve the maximum amount of resource extraction without depleting the resource from one harvest to the next. In forestry, MSY argues for cutting trees shortly after they go through their fastest stage of growth (at intermediate age), maximizing timber production over time. However, forest ecology is altered and species that depend on mature tree habitats are eliminated.

- **Ecosystem-based management** attempts to manage the harvesting of resources in ways that minimize impact on the ecosystem and ecological processes. This is a more challenging approach to implement and allows succession to occur in some areas of the managed land.

- **Sustainable forestry** is gaining ground and **sustainable forest certification** (by the Forest Stewardship Council, FSC) can now ensure the reliability and authenticity of these forest products by a "chain-of-custody certification" (meaning that all the steps from harvest to production have met strict standards). Specific attributes of FSC sustainable forestry include:

  1) protection of rare species and sensitive areas,
  2) safeguarding of water resources from pollution and disease,
  3) control of soil erosion, and
  4) minimal use of pesticides.

- More than 6% of the world's forests managed for timber production are now FSC certified.

- **Adaptive management** involves systematically testing different approaches and aiming to improve methods through time. This approach entails routinely monitoring the results of one's practices and adjusting them as needed.

- **The U.S. Forest Service**, established in 1905, set aside a **national forest system**, which today covers over 8% of U.S. land area. This is public land set aside to grow trees, produce timber, protect water quality and wildlife habitat, and also provide recreational opportunities in a **multiple-use** and **maximum sustainable yield** approach. In reality, however, timber production/harvest has most often been the primary use.

- In 1976, the **National Forest Management Act** was passed, mandating that every national forest draw up plans for renewable resource management with more explicit emphasis placed on multiple use and maximum sustainable yield. In this **new forestry** approach, timber harvests have become more ecosystem-based, with attempts to recover plant and animal communities that have been lost or degraded in the past.

- The vast majority of timber harvesting in the United States today takes place on *private* land owned by the timber industry and on the land owned by small private landowners.
- The U.S. timber industry focuses on **plantation forestry** in the South, where fast-growing tree species planted in **even-aged** monocultures dominate. This approach is quickly growing worldwide; however, ecologists view plantations more as crop agriculture than as a functional forest because biodiversity is lacking and susceptibility to the outbreak of pest species is common.
- Tree harvesting methods in today's forestry management plans include the following.

  - **Clear-cutting** is the most cost-effective harvesting method, but it has the greatest ecological impact because it involves cutting all the trees in an area, leaving only stumps.
  - In the **seed-tree approach**, small numbers of mature and vigorous seed-producing trees are left standing so they can reseed the logged area.
  - In the **shelterwood approach**, small numbers of mature trees are left in place to provide shelter for seedlings as they grow.
  - A **selection system approach** allows uneven-aged stand management because only some trees are cut at any one time; it is an approach unpopular with timber companies (more expensive) and loggers (more dangerous).
  - **Salvage logging** is the harvesting of trees for timber *after* a fire has burned a forest. A debate has arisen over whether the burned wood is more valuable left in place—for erosion control, wildlife habitat, and increased organic matter for soil. Recent studies have shown that salvage logging increases the risk of future severe fires.

- Past U.S. **fire policy** has been controversial, but scientists agree that there is a need to address the negative impacts of a century of fire suppression because many species and healthy ecosystems depend on fire. Catastrophic fires have resulted in the **wildland-urban interface** as a result of past fire suppression and increasing residential development along the edges of forested land.
- **Prescribed or controlled burns** clear away fuel loads, nourish the soil with ash, and encourage the vigorous growth of new vegetation. In the wake of major fires in California in 2003, **The Healthy Forests Restoration Act** was passed. This act encourages prescribed burning; however, timber companies have mainly focused on the physical removal of small trees, underbrush, and dead trees. Negative effects of soil erosion and a lack of forest regeneration have often been experienced.
- **Global climate change** is now worsening wildfire risk by bringing warmer weather to North America and drier weather to western U.S. states. Fire also appears to be encouraging certain insect pests,

particularly bark beetles, which are attracted to weakened trees. Millions of acres of western forest have now been devastated, killing billions of conifer trees and leaving them as fuel for fires. As climate change interacts with pests, diseases, and management strategies, U.S. forests are being profoundly altered, with the original moist forests being replaced by drier woodlands, shrublands, or grasslands.

### The major federal land management agencies and the lands they manage and protect

- The **U.S. Forest Service** manages all federally owned **national forests**. This agency falls within the Department of Agriculture.
- The **National Park Service** (1916) manages all the **national parks** and seashores, monuments, historic sites, national recreation areas, and national wild and scenic rivers within the Department of the Interior.
- The **U.S. Fish and Wildlife Service** administers the **national wildlife refuges**, begun in 1903, and protects habitat for wildlife while also encouraging hunting and fishing. This agency falls within the Department of the Interior.
- The **Bureau of Land Management (BLM)** is the nation's largest landowner, managing 260 million acres spreading across 12 western states and consisting primarily of rangelands used for livestock grazing.

### The various types of protected land designation, including evaluation of issues involved in their design

- National parks, national forests, national wildlife refuges, and wilderness areas form the central core of federally protected lands with an environmental connection in the United States.
- **Wilderness areas** are undeveloped lands initially set aside as a result of the Wilderness Act (1964), focusing on the preservation of undisturbed habitat, off-limits to development but open to hiking and other low-impact recreation. Wilderness areas have been established within national forests, national parks, and national wildlife refuges, and they are managed by the Bureau of Land Management (BLM).
- **Land trusts** are local or regional private nonprofit organizations that purchase land to preserve it in its natural condition. The Nature Conservancy, the world's largest land trust, uses science to select areas and ecosystems in the greatest need of protection.
- The **wise-use movement**, dedicated to protecting private property rights, opposes government regulation and promotes the transfer of federal lands to state and local control. This movement represents opposition to U.S. land protection policies and is especially prevalent in western states.
- Today 12.7% of the world's land area is designated for preservation in various types of parks, reserves, and protected areas. **Biosphere reserves** are tracts of land with exceptional biodiversity that couple preservation with sustainable development to benefit local people and are partly

managed internationally by the United Nations (the Maya Biosphere Reserve, Guatemala, is an example). **World Heritage Sites** are another type of international protected area and are managed for their natural or cultural value (Serengeti National Park in Tanzania is an example).

▌ **Habitat fragmentation** is a central issue affecting wildlife in their protected areas today. Expanding agriculture, cities, highways, and other impacts routinely divide large continuous expanses of habitat into small, disconnected ones in which many species cannot live. **Edge effects**, impacts that result because conditions along a fragment's edge are different from interior conditions, often are responsible for the failure of a species to breed or for increased predator pressure.

▌ A 2015 research study has concluded that 70% of global forests now lie within 1 km (0.62 mi) of an edge, indicating the increased potential for the direct influence of humans, the incursion of non-forest species, and the modification of forest microclimates.

▌ **Island biogeography theory**, developed by E. O. Wilson and Robert MacArthur in 1963, explains how species come to be distributed in "habitat islands." The theory predicts an "island's" species richness, involving immigration and extinction rates, based on the "island's" size and its distance from the larger/main reservoir of the species located in the unfragmented environment.

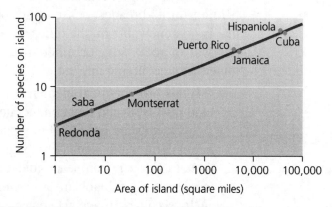

**Figure 12.27**   Island Biogeography Theory: Effect of Island Size on Species Richness

▌ Today, the *size* and *placement* in the landscape of protected areas are central issues in biodiversity conservation and protected area designation. Nicknamed the **SLOSS dilemma** ("single large or several small"), the question is whether it is better to make reserves large in size and few in number, or many in number but small in size. Wildlife corridors (protected land connecting fragmented "islands") may be one solution.

▌ In spite of worldwide efforts to increase the number of designated protected areas, global climate change now threatens to undo this investment. Species we had hoped to protect in parks may, in a warming world, become trapped in them. High elevation and high latitude species are most at risk from climate change because there is nowhere for them to go once a mountaintop becomes too warm or when an ice floe melts.

# Chapter 23: Minerals and Mining

---

**YOU MUST KNOW**

- The main types of mineral resources and their uses.
- The major mining methods, including a specific mineral resource example for each.
- Specific examples of the environmental and social impacts of mining.
- Examples of specific reclamation efforts and U.S. mining policy.
- Examples of the sustainable use of mineral resources.

---

### The main types of mineral resources and their uses

▮ **Minerals** are naturally occurring solid elements or inorganic compounds with a crystal structure, specific chemical composition, and distinct physical properties.

▮ **Mining**, in its broadest sense, describes the extraction of any resource that is nonrenewable on the timescale of our society. It is the systematic removal of rock, soil, or other material for the purpose of extracting minerals of economic interest.

   ▪ **Metals** are chemical elements that are typically lustrous, opaque, and malleable and that conduct heat and electricity.

   ▪ Most metals are not found in a pure state in Earth's crust but instead are present within **ore**, a grouping of minerals from which we extract metals. Copper, iron, lead, gold, and aluminum are among the many economically valuable metals we extract from mined ore.

   ▪ **Alloys** can be made when we mix, melt, and fuse a metal with another metal or a nonmetal, making a product that has the beneficial characteristics of each separate substance.

▮ The minerals we use come from all over the world. Some of the most economically important minerals (mostly metals), together with their major use and main nation of origin, include the following.

   ▪ *Titanium*, used for airplanes, from China
   ▪ *Copper*, used for electrical wiring, from Chile
   ▪ *Nickel*, used in alloy for stainless steel, from Cuba
   ▪ *Tin*, used for steel plating, from China

- *Tungsten*, used in metalworking, from China
- *Indium*, used for LCDs and solar cells, from Canada
- *Gold*, used as ingots for monetary value, from South Africa
- *Uranium*, used in nuclear fuel, from Australia
- *Tantalum*, extracted from coltan ore and used in capacitors and other electronic equipment, notably cell phones, from Australia and D.R. Congo

- Processing minerals exerts environmental impacts that are water-intensive, are energy intensive, and often cause a wide variety of air pollution problems. In addition, soil and water commonly become polluted by **tailings**, the portion of ore left over after metals have been extracted.
- **Nonmetallic minerals** and fuels are also mined: sand, gravel, phosphates, limestone, gemstones, uranium ore, coal, and other fossil fuels (petroleum and natural gas).

### *The major mining methods, including a specific mineral resource example for each*

- In 2015, raw materials from mining contributed $78 billion to the U.S. economy and employed more than 1.2 million people.
- **Strip mining** removes surface layers of soil and rock (the **overburden**) to expose the mineral resource below. Coal and oil sands are mined in this manner. By completely removing vegetative cover and nutrient-rich topsoil, strip mining obliterates natural communities over large areas and also produces **acid drainage** (when sulfide minerals in newly exposed rock surfaces react with oxygen and rainwater, producing sulfuric acid), a serious water pollution condition that further compromises aquatic life.
- **Subsurface mining** involves constructing deep shafts and tunnel systems when a resource occurs in concentrated pockets or seams deep underground. Zinc, lead, gold, copper, diamonds, and coal are mined in this manner. Subsurface mining is the most dangerous form of mining and produces a wide array of human health/respiratory hazards.
- **Open pit mining** is used when a mineral is spread widely and evenly throughout a rock formation or when the earth is unsuitable for tunneling. This involves digging a gigantic hole and removing the desired ore along with waste rock that surrounds the ore, leaving it in massive heaps outside the pit. Copper, iron, gold, and diamonds are mined in this manner.
- **Placer mining** uses running water to isolate minerals. Coltan ore, which contains the mineral tantalum (highly heat resistant and readily conducts electricity, making it ideal for capacitors used in all cell phones), is mined in this manner.
- **Mountaintop removal mining** involves removing several hundred vertical feet of a mountaintop after forests are clear-cut, topsoil removed, and rock blasted away to expose the mineral resource. Low-sulfur coal

in the Appalachian Mountains of the eastern United States is mined in this manner. Overburden is placed back onto the mountaintop, but the waste rock is unstable and typically is dumped into adjacent valleys.

- **Solution mining** dissolves and extracts resources in place and is often used for mining salts, some uranium, and copper.
- **Ocean mining** can involve the extraction of some minerals from seawater (magnesium) and/or mining the ocean floor for sand, gravel, and especially phosphate deposits used in fertilizers.

### *Specific examples of the environmental and social impacts of mining*

- Many mining methods completely remove vegetation and topsoil, destroying the biological community and habitat for many species: strip mining, open pit mining, and mountaintop mining are the leading offenders.
- **Acid drainage** occurs when water leaches compounds (especially sulfide minerals) from freshly exposed waste rock, leading to water pollution that is highly toxic (due to its low pH) to aquatic organisms. The Berkeley Pit, a former open pit copper mine in Montana, is today one of the largest Superfund toxic waste cleanup sites because of the highly acidic groundwater pollution it caused.
- When overburden (from strip mining, open pit mining, and mountaintop mining) is replaced, it is highly unstable and often erodes because the original vegetation has been removed. As a result, high levels of sediment are deposited in surrounding waterways.
- **Mountaintop removal** for coal destroys forests, mountaintop habitats, and adjacent valleys and streams because the waste rock is most often dumped in these areas. Loss of biodiversity and loss of ecosystem services are the most obvious impacts; however, loss of rare, endemic species and loss of the ecosystem's ability to cycle nutrients and produce organic matter for downstream systems are less apparent but just as vital.
- Some minerals exposed from the mining processes, such as selenium, are known to **bioaccumulate** in organisms, causing birth defects.
- Various studies have found that when over 5–10% of a watershed's area is disturbed by human alteration, the water quality and biodiversity in a stream decline. Mountaintop mining typically disturbs percentages ranging from 17–51%.
- Mining often has negative health impacts on miners, especially with regard to respiratory conditions (e.g., black lung and emphysema).
- Mining often causes adverse living conditions for people living near mines due to increased noise pollution, groundwater pollution, and inhaling air pollutants and dust. Increased flood damage, due to flash flooding potential (resulting from the removal of vegetation and topsoil, which alters the topography), is also a major concern.

### *Examples of specific reclamation efforts and U.S. mining policy*

▌ Because of the environmental impacts of mining, the United States and many other developed nations now require that mining companies restore or reclaim surface-mined sites following mining. The **Surface Mining Control and Reclamation Act** (1977) mandates restoration efforts, requiring companies to post bonds to cover reclamation costs before mining can be approved.

▌ Mine reclamation efforts are challenging and often not able to restore the original ecology because of toxic soil and water conditions left behind by the mining operation.

▌ Mining policy in the U.S. is still being dictated by the **General Mining Act of 1872**. This act encourages prospecting for minerals on federally owned land by allowing claims to be staked on any plot of public land open to mining. No payments back to the public are required, and only recently has restoration of the land after mining been mandated.

▌ A 2014 study has found that there may be a potential for treating fracking wastewater with acid mine drainage water, resulting in the removal of some contaminants (barium, strontium, sulfate, and radioactive radium) as precipitates (strontium barite), producing wastewater less damaging to the environment. Additional research into this possible approach is ongoing.

▌ Superfund sites, like Montana's Berkeley Pit, are slowly being reclaimed; however, the number of designated Superfund sites far exceeds those being actively cleaned up.

### *Examples of the sustainable use of mineral resources*

▌ By encouraging **recycling of mineral resources**, we can address the two major challenges of their finite supply and the environmental damage issues caused by their mining. For example, currently 33% of our copper and 80% of our lead come from recycled sources such as pipes and wires. Over 40% of the aluminum in the U.S. is recycled, which is a good thing because it takes over 20 times more energy to extract virgin aluminum from ore than it does to obtain it from recycled sources. Overall, 35% of all mineral resources are currently recycled in the U.S.

▌ Several factors now affect how long a given mineral resource will last: discovery of new reserves, development of new extraction technologies, changing consumption patterns, and the extent of recycling of a specific mineral.

> **Electronic waste**, or e-waste, from discarded cell phones, computers, and other electronic products is rising fast and contains many valuable minerals, such as tantalum. Recycling, refurbishing, and reusing these devices rather than discarding them, is a more sustainable approach and will require more efforts and focus. Currently, only about 10% of old cell phones are recycled.

**Table 23.1** Recycled minerals in the United States

| Mineral | U.S. Recycling Rate |
|---|---|
| Gold | Slightly less is recycled than is consumed |
| Iron and steel scrap | 85% for autos, 82% for appliances, 72–98% for construction materials, 70% for cans |
| Lead | 69% consumed comes from recycled post-consumer items |
| Tungsten | 59% consumed is from recycled scrap |
| Nickel | 45% consumed is from recycled nickel |
| Zinc | 37% produced is recovered, mostly from recycled materials used in processing |
| Chromium | 34% is recycled in stainless steel production |
| Copper | 32% of U.S. supply comes from various recycled sources |
| Aluminum | 30% produced comes from recycled post-consumer items |
| Tin | 30% consumed is from recycled tin |
| Germanium | 30% consumed worldwide is recycled. Optical device manufacturing recycles more than 60% |
| Molybdenum | About 30% gets recycled as part of steel scrap that is recycled |
| Cobalt | 28% consumed comes from recycled scrap |
| Niobium (columbium) | Perhaps 20% gets recycled as part of steel scrap that is recycled |
| Silver | 15% consumed is from recycled silver. U.S. recovers as much as it produces |
| Bismuth | All scrap metal containing bismuth is recycled, providing less than 10% of consumption |
| Diamond (industrial) | 7% of production is from recycled diamond dust, grit, and stone |

Data are for 2015, from U.S. Geological Survey, 2016. *Mineral commodity summaries 2016*. Reston, VA: USGS.

▌ Minerals are nonrenewable resources and many are currently limited in supply. For example, the dwindling supplies of indium (mined in Canada and a previously obscure metal) are now widely used for LCD screens and solar cells, and are projected to last only another 30+ years. A lack of indium threatens the production of high-efficiency photovoltaic solar cells. Also, of the 1.7 billion cell phones sold each year, only 10% are recycled. Recycling old electronic devices will help conserve valuable minerals such as tantalum (from coltan ore mined in the Democratic Republic of Congo), and also keeps the growing volume of e-waste out of landfills.

*Multiple-Choice Questions*

1. The soil horizon commonly known as subsoil is the
   (A) A horizon.
   (B) O horizon.
   (C) B horizon.
   (D) C horizon.
   (E) E horizon.

2. Which of the following is the *primary* environmental advantage of no-till agriculture?
   (A) The concentration of essential $CO_2$ in the soil is increased.
   (B) The crop residues reduce the O horizon.
   (C) Migratory bird populations are undisturbed.
   (D) The use of herbicides improves the overall stability of the soil.
   (E) The undisturbed soil is less susceptible to erosion.

3. Genetically modified (GM) organisms are best described as those that
   (A) have genes of several different members of the same species.
   (B) have genes of several members of a different species.
   (C) have genes that have been artificially created in a laboratory.
   (D) have genes that have bacterial genes for increased resistance, plus the most desirable genes of the species.
   (E) have genes that code for desirable traits from one species that are modified and neutralized so they can then be inserted into another species, thus displaying the desired trait.

4. Which of the following best explains how a pest develops resistance to a chemical pesticide?
   (A) The pest develops adaptations to deal with adverse effects in its environment.
   (B) Mutation and genetic recombination occur naturally.
   (C) Genetically modified organisms are usually best adapted to be resistant.
   (D) Natural selection takes place.
   (E) Industrial agribusiness practices of using large amounts of herbicides encourage pest resistance.

5. Island biogeography theory, developed by E. O. Wilson and Robert MacArthur in 1963, has many implications for
   (A) developing maximum biodiversity on remote islands around the world.
   (B) protecting the biodiversity of endangered island ecosystems.
   (C) planning wise use of island resources.
   (D) developing national parks of adequate size and with sufficient wildlife corridors.
   (E) developing wilderness areas that will be able to sustain themselves for the long term.

6. Second-growth forests
   (A) are less abundant on Earth today than they were 200 years ago.
   (B) are forests that establish themselves after virgin timber has been removed.
   (C) are those forests in wilderness areas allowed to grow back after deforestation.
   (D) are forests whose timber has a second-rate economic value.
   (E) are located mostly in British Columbia and Alaska.

7. Many types of mining, such as those used to mine coal and copper, produce a specific type of water pollution called
   (A) eutrophication.
   (B) suspended limestone particles.
   (C) salinization.
   (D) sediment load.
   (E) acid drainage.

8. International seedbanks have been established for the *primary* purpose of
   (A) increasing demand and lowering prices for valuable plant products.
   (B) finding more available substitutes for scarce agricultural crops for developing countries.
   (C) increasing productivity in developed countries.
   (D) international cooperation.
   (E) preserving declining crop diversity.

9. The environmental consequence of overfertilization of agricultural lands includes
   (A) sustained fertile soils in years to come.
   (B) rapid spread of crops into nearby areas.
   (C) large crop yields per acre.
   (D) eutrophication in nearby waters.
   (E) very large fruits and vegetables.

10. As scientists continue to study forest fragmentation concerns, an important conclusion has been that
    (A) small forest fragments lose more species than large fragments.
    (B) large forest fragments contain more species and are more vulnerable to species loss.
    (C) forest fragments closer to one another lose more species overall.
    (D) climate change has neutralized forest fragmentation effects.
    (E) forest fragmentation has had minimal effects on biogeochemical cycles.

11. The Green Revolution is most closely associated with which time period?
    (A) the most recent decade, when industrial agribusiness has been at its peak
    (B) between World War I and World War II
    (C) between World War II and the present
    (D) between the early 1990s and the present
    (E) immediately after the Industrial Revolution, when food shortages were the greatest

12. Which of the following is an important feature of integrated pest management?
    (A) It requires subsistence agricultural practices, including composting.
    (B) It makes use of the natural enemies of pests.
    (C) It makes effective use of GMO-developed plants.
    (D) It makes effective use of *Bacillus thuringiensis*.
    (E) It involves relocating pests back to their native habitat, where they will do minimal damage.

13. Wilderness areas
    (A) allow hunting as long as proper permits are acquired.
    (B) are biosphere reserves managed by UNESCO.
    (C) were set up under the wise-use and maximum sustainable yield approach.
    (D) are off-limits to development but are open to low-impact recreation.
    (E) allow no roads, no development, no permanent structures, and no recreation.

14. The Healthy Forests Restoration Act
    (A) was passed by Congress with the intent of ensuring multiple use and sustainable yield of forest land.
    (B) restricts the use of recreational vehicles on national forest lands.
    (C) directs timber companies to remove small trees, underbrush, and dead trees to reduce fires in national forest lands.
    (D) sets appropriate guidelines for managing the health and development of national forest lands.
    (E) offers subsidies to timber companies when they replant trees in national forest land.

15. The U.S Surface Mining Control and Reclamation Act of 1977 requires
    (A) all mined minerals to be processed and sold within the United States.
    (B) minimal remediation of water pollution, except in the case of toxic uranium mining.
    (C) restoration of the identical ecosystem and biodiversity present prior to the mining activity.
    (D) mining companies to post bonds to cover restoration of the mined area before permits are granted.
    (E) 10% of sales revenues to be donated to national park maintenance.

16. IPM is the agricultural approach involving
    (A) biocontrol and alternative tillage methods.
    (B) treatment with chlorine and/or ozone.
    (C) pollinators, transgenic crops, and CAFOs.
    (D) Roundup Ready crops.
    (E) Green Revolution techniques.

17. Leaching
    (A) can help plant growth if done sustainably.
    (B) is caused by movement of water upward through soil from the water table.
    (C) adds nutrients to soil naturally from weathered parent material.
    (D) removes water-soluble nutrients from soil.
    (E) is a common agricultural practice that improves the sustainability of the soil.

18. Humus is
    (A) the artificial fertilizer applied to monocultures.
    (B) composed of organic compounds and increases the water-holding capacity of soils.
    (C) a layer sometimes found in a soil horizon that is created by eluviation.
    (D) one of the primary causes of desertification if present in excess.
    (E) promotes eutrophication when agricultural practices allow it to get into waterways.

19. Crops genetically engineered to fix nitrogen would have which of the following advantages?
    (A) increased disease resistance
    (B) increased tolerance of nitrous oxides in the air
    (C) decreased absorption of toxic chemicals
    (D) decreased need for chemical fertilizer
    (E) overall increased tolerance of herbicides and pesticides

20. Critics of the Green Revolution most often cite
    (A) the overabundance of food produced, leading to unusually low food prices.
    (B) the necessity for seed banks, now that we are using so many varied seeds to grow crops worldwide.
    (C) the increased practice of monocropping, leading to increased risk of overall crop failure without the widespread use of pesticides.
    (D) new strains of GMO seeds that have transgenes, which may escape into nature, causing unknown and possibly harmful effects.
    (E) new strains of GMO seeds that might cause human health problems.

**Use Figure 2 below to answer questions 21 and 22.**

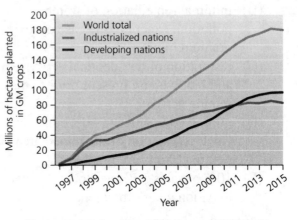

**Figure 2** Organic versus Conventional Farming Comparisons

21. A major conclusion that could be reached from the above data is
    (A) greenhouse gas emission is improved with conventional farming methods.
    (B) energy input is reduced with organic farming methods.
    (C) profit comparisons are negligible between these two farming methods.

(D) yields of conventional farming are high compared to organic farming methods.
(E) humans prefer organic foods for their healthful benefits.

22. When comparing greenhouse gas emissions, conventional farming
    (A) produces less emissions than organic farming.
    (B) produces approximately 1000 lb $CO_2$/acre/yr more.
    (C) produces approximately 1000 lb $CO_2$/acre/yr less.
    (D) produces approximately 500 lb $CO_2$/acre/yr more.
    (E) produces approximately 50% more emissions than organic farming.

**Use Figure 10.23 below to answer questions 23 and 24.**

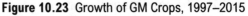

**Figure 10.23** Growth of GM Crops, 1997–2015

23. The percent change in the growth of the world's total GM crops between 2009–2012 is approximately
    (A) 170%.
    (B) 80%.
    (C) 40%.
    (D) 30%.
    (E) 20%.

24. The graph evidence provides support for the conclusion that in the last 5 years, GM crops have been
    (A) growing faster in industrialized nations.
    (B) growing faster in developing nations.
    (C) positively influencing world food supply.
    (D) negatively influencing world food supply.
    (E) providing over 50% of the world's food supply.

25. Conservation tillage is an agricultural practice that
    (A) maximizes intact ground cover to prevent soil erosion.
    (B) conserves all native grasses in order to promote intact ecosystems.
    (C) removes all weeds and competing plants in order to maximize crop yields.
    (D) promotes polyculture, including cattle grazing, to naturally fertilize fields.
    (E) reduces the effects of climate change and desertification.

26. Forests provide many ecosystem services; however, the one that is receiving the most focus for climate change scientists is
    (A) protecting watersheds.
    (B) carbon sequestration.
    (C) habitat protection.
    (D) CFC neutralization.
    (E) methane storage.

27. The major environmental problem arising from feedlots is
    (A) damaged soil that will never be able to grow food crops.
    (B) pollution of surface and groundwater, resulting in eutrophication.
    (C) climate change arising from increased carbon dioxide production.
    (D) increased use of antibiotics and increasing livestock resistance to these antibiotics.
    (E) increased incidences of disease such as Lyme disease and botulism in livestock populations.

28. Sustainable forest certification involves
    (A) a shelterwood approach to forest management.
    (B) controlled burns on a regular basis.
    (C) protecting all secondary growth forests.
    (D) wildlife corridors developed in urban landscapes.
    (E) a chain of custody following for all harvested material.

## Free-Response Question

The *San Francisco Buzz* recently published an article—"Is Our Food Supply Safe for Today and the Future?"— that received much attention across the U.S. Read the following brief excerpt and answer the questions below.

> Much of our food supply today is the result of genetic engineering of plants. For thousands of years, humans have been influencing the genetic make-up of our crop plants to increase food production. However, now the genes of plants are being altered in new ways to make crops more pest-, drought- and cold-resistant, resulting in genetically modified organisms (GMOs). Many citizens have begun to worry about the health impacts of eating GM foods, although the scientific community is more concerned about the potential ecological effects. This concern difference has sparked a vigorous debate on the future path of food production and global food security.

a) Describe two significant *differences* in the techniques involved in producing GMOs, as compared to the traditional agricultural techniques used in the past to improve crop pesticide resistance.

b) i. Identify two important GMOs on the market today.
   ii. For each of your GMO choices above, explain one *environmental* or *health benefit*.
   iii. For one of these GMO choices, describe an *ecological concern* expressed by the scientific community.

c) So far, GM crops have not lived up to their promise of feeding the world's hungry. Explain why this is true.

# The Urban Environment

## *Chapter 13: The Urban Environment: Creating Sustainable Cities*

---

> **YOU MUST KNOW**
>
> - The current trends and scale of urbanization.
> - The causes and effects of urban sprawl.
> - The environmental impacts and advantages of urban centers.
> - Specific approaches and policies that promote sustainable cities.
> - The transportation options and the characteristics of urban parks and green buildings.

### *The current trends and scale of urbanization*

▌ **Urbanization**, the shift from living in rural areas to cities and suburbs, may be the single greatest change our society has undergone since its transition from nomadic hunter-gatherer to a sedentary agricultural lifestyle.

▌ Worldwide, the proportion of population that is urban rose from 30% (in 1960s) to more than 50% today. In the United States, this urban percentage passed 50% shortly before 1920 and now stands at 80%, with approximately 50% in suburbs.

▌ In developing countries, most of the population still resides on farms; however, most of the fast-growing cities today are in the developing world (Delhi, India; Lagos, Nigeria; and Karachi, Pakistan), where population growth exceeds economic growth, producing conditions of overcrowding, pollution, and poverty as these countries continue to industrialize.

▌ U.S. cities grew rapidly throughout the 19th and early 20th centuries as a result of immigration and increased industrialization. By the 1950s, the rise in the number of automobiles, expanding road networks/interstate highways, and inexpensive and abundant oil were factors enabling people to move to suburbs and away from inner cities. Twenty-first century technology (cell phones and the Internet) has reinforced the spread of urban populations to suburban areas as they allow people to communicate more easily from locations away from major urban centers.

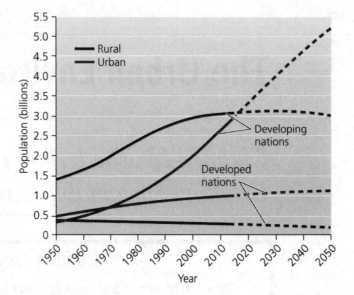

**Figure 13.1**   Urban Growth in Developing vs. Developed Nations Today

### The causes and effects of urban sprawl

▌ **Urban sprawl** is the spread of low-density urban or suburban development outward from an urban center, removing clear boundaries between the two. The landscape is often characterized by "strip malls," cluster housing, office parks, and "big box" retail stores separated by miles of roads.

▌ Many urban researchers define *sprawl* as the physical spread of development at a rate that exceeds the rate of population growth.

▌ Worldwide, the main cause of urban sprawl is increased human population growth. There are simply more people alive each year than the previous year.

▌ The main causes of urban sprawl in the United States include increased human population along with the following important factors.

  ▪ The development of federal and state **interstate highway systems** in the 1950s and 1960s promoted living in suburbs while continuing to work in cities. Also, **telecommunications** and the **Internet** have fostered movement away from urban centers.

  ▪ **Living costs** are often less in suburban and rural areas, as one can obtain more land and a larger house for much less money.

  ▪ **Urban blight** results when people move away from a city to the suburbs, causing the city's tax revenue to decline, which in turn forces city services to decline, encouraging even further flight from cities.

  ▪ Government policies, such as zoning laws, have affected where people have chosen to live. **Zoning** is a planning tool often used to separate industry and business from residential neighborhoods. Governments that use zoning can classify land areas into zones in which certain land uses are prohibited, such as the building of a factory in a residential area, sometimes restricting the use of private land for the greater good of the community.

▌ Since sprawl means different things to different people (some positive, some negative), findings by urban researchers on the effects of sprawl are an important tool. Most studies show that urban sprawl:

- **constrains transportation options**, essentially forcing people to own more cars and drive more miles each day. During the 1980s and 1990s, the average length of work trips rose by 36%, and total vehicle miles driven rose three times faster than population growth.

- **increases dependence on nonrenewable petroleum** with its many economic and environmental consequences, particularly those associated with climate change (increased $CO_2$ emissions from cars).

- **drains tax dollars** by diverting money that could be spent on maintaining and improving downtown centers because of the need to extend road, water, sewage, and communications systems outward into suburbs.

- **compromises air and water quality**. With increased reliance on automobile use comes increased carbon dioxide emissions, as well as nitrogen and sulfur-containing air pollutants, which in turn lead to tropospheric ozone, urban smog, acid precipitation, and global climate change. Motor oil and road salts can pollute waterways.

- **contributes to increased land conversion** from existing forests, fields, and farmland. These forested and agricultural lands provide vital resources (food, recreation, wildlife habitat) and ecosystem services (water purification, climate modification, etc.).

- **compromises human health**, in addition to the air and water pollution issues. People rely on cars to do their daily errands instead of walking, and the result is a decrease in physical activity.

## The environmental impacts and advantages of urban centers

▌ **Sustainable cities** are cities that can function effectively and prosperously over the long term, providing generations of residents a good quality of life far into the future. Portland, Oregon, is a good example of a sustainable city that is enhancing the quality of life for its citizens and protecting environmental quality at the same time.

▌ Cities and towns are *sinks* for resources. High per capita resource consumption creates substantial waste and pollution in the long-distance transportation of these resources. However, if all the world's 3.9 billion urban residents were instead spread evenly across the landscape, there would *likely be more transportation* required to provide people the same access to resources and goods.

- Cities minimize per capita consumption costs through the efficient delivery of goods and services because people are densely concentrated.
- Urban citizens often suffer **noise and light pollution** (undesired ambient sound and light), a constant aspect of city life.
- **Urban heat island** effects result from the concentration of heat-generating buildings, vehicles, and factories, causing thermal pollution. Also, the dark surfaces of asphalt pavement and heat-absorbing buildings disrupt the normal convective cooling that would otherwise occur.
- Urbanization preserves land, the important idea behind **urban growth boundaries (UGB)**. For example, Oregon's law requires every city and county to draw up a comprehensive land use plan, incorporating UGBs, with an intent to revitalize city centers, prevent suburban sprawl, and protect farmland, forests, and open landscapes.
- **Urban ecology** supports the central idea that cities should be viewed as ecosystems and that the fundamentals of ecosystem ecology and systems science apply to the urban environment. Major urban ecology projects are ongoing in Baltimore, MD, and Phoenix, AZ, as researchers are studying these cities with a view toward their potential as ecological systems. Using this approach, urban sustainability advocates suggest that cities should strive to:
  - Maximize efficient use of resources.
  - Recycle as much as possible.
  - Account for external costs.
  - Use locally produced resources.
  - Encourage urban agriculture.
  - Offer tax incentives to encourage sustainable practices.

### *Specific approaches and policies that promote sustainable cities*

- **Smart growth** focuses on strategies that encourage the development of sustainable, healthy communities.
  - This approach includes use of urban growth boundaries (UGBs).
  - Smart growth means "building up, not out," focusing development and economic investment in existing urban centers and favoring multi-story shop-houses and high-rises.
  - The EPA lists 10 basic principles of smart growth, with some of the more important features being mixed land uses (residential, recreation, and business lands together), walkable neighborhoods, an economic range of housing opportunities, and preserving open space.

**TABLE 13.1** Ten Principles of "Smart Growth"

▶ Mix land uses

▶ Take advantage of compact building design

▶ Create a range of housing opportunities and choices

▶ Create walkable neighborhoods

▶ Foster distinctive, attractive communities with a strong sense of place

▶ Preserve open space, farmland, natural beauty, and critical environmental areas

▶ Strengthen existing communities, and direct development toward them

▶ Provide a variety of transportation choices

▶ Make development decisions predictable, fair, and cost-effective

▶ Encourage community and stakeholder collaboration in development decisions

*Source:* U.S. Environmental Protection Agency

### *The transportation options and the characteristics of urban parks and green buildings*

▍ **Transit-oriented development** is critical to new urbanism approaches which seek to design neighborhoods on a walkable scale with homes, businesses, and other amenities all close together for convenience.

▪ Public buses, trains, subways (heavy rail), and light rail (smaller rail systems powered by electricity) are all options. Portland's light rail systems are currently carrying 100 million riders per year.

▪ Rail systems have been calculated to save taxpayers a total of $67.7 billion in costs for congestion, consumer transportation, parking, and road maintenance—far more than the $12.5 billion that governments currently spend to subsidize rail systems per year. New York City's subways, Washington D.C.'s Metro, and San Francisco Bay Area's BART carry more than one-fourth of each city's daily commuters.

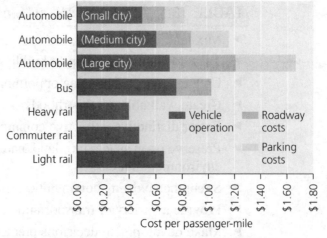

**(c) Operating costs for different modes of transit**

**Figure 13.10c**   Operating Costs for Different Modes of Transit

- The United States has neglected mass transit for the past several decades, instead choosing to invest in road networks for cars and trucks.

■ **Parks and open spaces** are key elements of sustainable cities. America's city parks began to arise in the late 19th century with the establishment of New York City's Central Park, designed by the leading American landscape architect, Frederick Law Olmsted. Boston, Philadelphia, and San Francisco subsequently developed major city parks.

■ **Greenways**, strips of land that connect parks or neighborhoods, often run along rivers, streams, or canals and provide access to networks of walking trails. The newly developed High Line Park in New York City is used by more than 13,000 people per day.

■ **Greenbelts** are long and wide corridors of parklands or open space, often encircling an entire urban area.

■ **Green buildings**, structures that incorporate various means of reducing the ecological footprint of a building's construction and operation, are a key approach toward sustainable cities. They are built from sustainable materials, minimize their use of energy and water, minimize health impacts on their occupants, limit their pollution, and recycle their wastes.

- The **Leadership in Energy and Environmental Design (LEED)** program is a certification program for green buildings.
- Green building techniques add expense to construction, but the added cost is generally less than 3–10%.
- LEED certification is booming throughout the United States, with Portland, Oregon, leading the way.

■ New York City unveiled an extensive sustainability program in 2007, hoping to make it "the first environmentally sustainable 21st-century city." Reduction of greenhouse gas emissions, improved mass transit, increased tree planting, and a proposal to charge drivers for driving into downtown Manhattan are a few key approaches of this sustainable city's program. Recently, this program has been renamed One NYC and is adding new dimensions to address economic equity.

**TABLE 13.2** Green Building Approaches for LEED Certification

| Approaches That Are Rewarded | Maximum Points |
|---|---|
| ENERGY: Monitor energy use; use efficient design, construction, appliances, systems, and lighting; use clean, renewable energy sources | 37 |
| THE SITE: Build on previously developed land; minimize erosion, runoff, and water pollution; use regionally appropriate landscaping; integrate with transportation options | 21 |
| INDOORS: Improve indoor air quality, provide natural daylight and views; improve acoustics | 17 |
| MATERIALS: Use local or sustainably grown, harvested, and produced products; reduce, reuse, and recycle waste | 14 |
| WATER USE: Use efficient appliances inside; landscape for water conservation outside | 11 |
| **Points above may total up to 100. Then, up to 10 bonus points may be awarded for:** | |
| INNOVATION: New and innovative technologies and strategies to go beyond LEED requirements | 6 |
| THE REGION: Addressing environmental concerns most important for one's region | 4 |

*Out of 110 possible points, 40 are required for LEED certification, 50 for silver, 60 for gold, and 80 for platinum levels. Point allocation shown is for new construction. Allocation varies slightly for renovations and specific types of buildings.*

# Chapter 14: Environmental Health and Toxicology

## YOU MUST KNOW

- The major goals of environmental health including the major environmental health hazards.
- The major types of toxic substances in the environment, the factors affecting their toxicity, and the defenses that organisms possess against them.
- How toxic substances move in the environment and how they affect organisms and ecosystems.
- How toxic substances are studied in the laboratory and in the environment.
- How to evaluate risk assessment and risk management, including different philosophical approaches to risk.
- How environmental health hazards are regulated in the United States and in the international community.

### The major goals of environmental health including the major environmental health hazards

▌ The study and practice of **environmental health** assesses environmental factors that influence our health and quality of life. Scientific studies can identify the necessary steps needed to minimize the risks of encountering hazards and to mitigate the impacts of the hazards we do encounter.

▌ Environmental health hazards can be natural or **anthropogenic** (human-caused), can occur outdoors or indoors, and are categorized into four main types: biological, chemical, physical, and cultural.

 ▪ **Biological hazards** cause the most human deaths. **Infectious diseases** (diseases caused by pathogens) result from viruses, bacteria, protists, fungi, parasitic worms or other pathogens or vectors (a **vector** is an organism that transfers the pathogen to the host, for example, the mosquito that transmits the malaria-causing protist *Plasmodium*). The spread of infectious disease today is much easier than in the past because of the increased density of human populations; extensive travel; the emergence of new strains of diseases (Ebola in West Africa in 2014); and global climate change, which causes movement of tropical diseases to new locations.

 ▪ **Chemical hazards** include the many synthetic chemicals that society manufactures, such as pharmaceuticals, pesticides, plastics, and industrial chemicals.

- **Physical hazards** arise from processes that occur naturally in our environment, such as ultraviolet (UV) radiation from sunlight.
- **Cultural hazards** result from our place of residence, socioeconomic status, occupation, or our behavioral choices. Cigarette smoking, drug use, and nutritional choices are examples in this category.

▌ **Indoor health hazards** are primarily due to chemicals and are important to consider because we spend approximately 90% of our lives indoors.

- **Cigarette smoke** and **radon** are both highly toxic. Radon is a colorless, odorless radioactive gas that seeps up from the ground in areas with certain types of bedrock, accumulating in basements and homes with poor air circulation.
- **Asbestos**, used in construction for its insulation, sound-proofing, and fire-retardant qualities, has been confirmed to cause lung cancer.
- **Lead poisoning** is a serious neurological health concern that can cause learning abnormalities, damage to the brain, liver, kidney, and stomach, and even death. Today, lead poisoning can come from lead-based paint exposure and drinking water that has passed through lead pipes.
- **Polybrominated diphenyl ethers (PBDEs)** are synthetically created compounds providing fire-retardant properties and used in a wide array of products including computers, televisions, plastics, and furniture. These chemicals appear to act as hormone disruptors and neurological toxins.
- **Bisphenol A (BPA)** is a synthetically created compound used in the plastics industry to create polycarbonate, a hard, clear type of plastic that can be used in water bottles, food containers, eating utensils, eyeglass lenses, CDs and DVDs, laptops, and other electronics. Scientific studies have recently categorized BPA as an **endocrine disruptor** because it mimics the female sex hormone estrogen.

▌ **Disease** is a major focus of environmental health, given that the major killers (cancer, heart disease, and respiratory disorders) are **noninfectious**—instead, they are developed through environmental factors (proximity to industrial centers, malnutrition, poverty, and poor hygiene caused by compromised water quality).

▌ The plague (caused by a bacterium carried by fleas), malaria (caused by a protist in the genus *Plasmodium* and carried by mosquitoes), tuberculosis (caused by a bacterium affecting lungs), and diarrheal diseases are infectious diseases that have affected humans for centuries.

▌ **Emergent infectious diseases** are defined as diseases that were previously not described or have not been common for at least several prior decades. Acquired immunodeficiency syndrome (AIDS), severe acute respiratory syndrome (SARS), Ebola hemorrhagic fever, West Nile virus, and swine flu (H1N1) are in this growing category. Many of these new diseases have come from viruses that normally infect animal hosts but have unexpectedly jumped to human hosts through mutation.

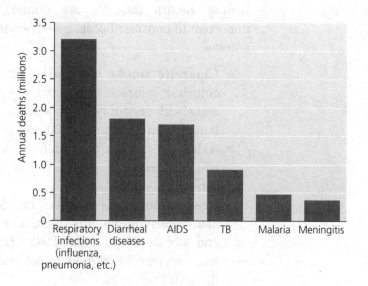

**(b) Leading causes of death by infectious diseases**

**Figure 14.3b**  Leading Causes of Death by Infectious Diseases

▌ It is thought that human-induced global warming and climate change are causing tropical diseases such as malaria, dengue, cholera, and yellow fever to begin expanding into temperate climate zones.

▌ Human land disturbance, which might create standing pools of water in formerly well-drained areas, can boost mosquito populations, causing renewed outbreaks of malaria that were formerly controlled by insecticides such as DDT.

▌ Improvements in **nutrition**, wider availability of **clean drinking water**, proper **sanitation**, and **education** are the four key approaches for lessening environmental health risks of both historical and emerging infectious diseases.

### The major types and distribution of toxic substances in the environment

▌ **Toxicology** is the science that examines the effects of poisonous substances on humans and other organisms. Environmental toxicology deals specifically with toxic substances that come from or are discharged into the environment and uses organisms other than humans as test subjects.

▌ **Toxins** are natural toxic chemicals manufactured in the tissues of living organisms (e.g., chemicals that plants use to ward off herbivores). In addition, we are exposed to many **synthetic** (artificial or human-made) chemicals.

▌ Rachel Carson's **Silent Spring** (1962) first brought to the public's attention the scientific studies and medical case histories regarding the insecticide DDT. Carson's message was that DDT and artificial pesticides are hazardous to people, wildlife, and ecosystems. The chemical industry challenged Carson's book vigorously; however, the use of DDT was banned in the United States in 1973 and is now illegal in many nations.

▌ A **"circle of poison"** continues to exist. Although the United States has banned the use of DDT, it continues to manufacture and export the compound to developing nations. Thus, it is possible that pesticide-laden food can be imported back into the United States.

▌ Toxic chemicals are classified into five major groups based on their particular effects on living organisms.

   ▪ **Carcinogens** are chemical substances or radiation that cause cancer. **Mutagens** are substances that cause damage to the genetic material of a cell. Some of the most well-known carcinogens include asbestos, radon, formaldehyde, and the chemicals (PAHs) found in cigarette smoke.

   ▪ **Neurotoxins** are chemicals that disrupt the nervous systems of animals. Many insecticides, lead, and mercury are examples.

   ▪ **Teratogens** are chemicals that interfere with the normal development of embryos. Thalidomide and alcohol are examples.

   ▪ **Allergens** are chemicals that overactivate the immune system, causing an immune response when one is not necessary, often producing severe breathing difficulties. For a few individuals, peanuts and some drugs such as penicillin produce hyperallergic reactions.

   ▪ **Endocrine disruptors** are toxicants that interfere with normal human hormonal balance. These chemicals can affect an animal's endocrine system by blocking the action of hormones or accelerating their breakdown. Effective in very low concentrations and not directly causing death, endocrine disruptors pose unusual challenges for toxicology. Biphenol A (BPA), which mimics estrogen, and the phthalates contained in many plastics and cosmetics are examples of endocrine disruptors. Theo Colburn's **Our Stolen Future** (1996) was the first study to propose the concept of endocrine disruption; however, debate has been generated by the chemical industry as a result of the potential economic threat which the removal of BPA could present.

### How toxic substances move in the environment and how they affect organisms and ecosystems

▌ Toxic substances are released from agriculture, industrial, and domestic activities and may concentrate in and move through surface water and groundwater, or they may travel long distances through the atmosphere.

▌ When affected species become smaller in number, their decline can, in turn, affect other species. These **cascading impacts** can cause changes in the composition of the biological community.

▌ Some chemicals degrade quickly and become harmless (such as the Bt toxin used in biocontrol and genetically modified crops); however, others may remain unaltered and persist for months, years, or decades (such as mercury, DDT, PCB).

▌ **Bioaccumulation** is the buildup of persistent chemicals in the tissues of an animal, producing a greater concentration of the substance than exists in the surrounding environment. Toxic substances that bioaccumulate (are persistent and fat soluble) may be transferred to other organisms when predators consume prey, a process known as **biomagnification**. Biomagnification means that the higher an organism is in the food web, the more toxicants it contains. A prime example has been the high concentration of the pesticide DDT in top predators, such as birds of prey. Bald eagles, peregrine falcon, brown pelican, and osprey have all suffered eggshell thinning because of high levels of DDT.

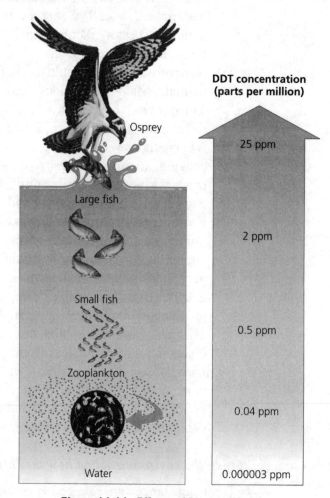

**Figure 14.14** Effects of Biomagnification

▌ Individuals vary in their responses to hazards. Fetuses, infants, and young tend to be much more sensitive to toxicants than adult organisms.

### *How toxic substances are studied in the laboratory and in the environment*

▐ Studies of wildlife can inform scientists about potential human health effects. Louis Guillette's studies of Florida alligators showed bizarre reproductive problems that were later attributed to agricultural runoff containing DDT and the herbicide atrazine. Tyrone Hayes has found similar reproductive problems in lab studies with frogs exposed to atrazine.

▐ **Epidemiological studies** are large-scale comparisons among groups of people, usually contrasting a group known to have been exposed to some hazard and a group that has not. These studies often last for long periods of time (years or decades). Asbestos mining and the Chernobyl nuclear disaster have both been analyzed from an epidemiological standpoint.

▐ To establish causation, manipulative experiments involving animal or plant studies are needed because subjecting humans to massive doses of toxic substances in lab experiments would be unethical.

  ▪ **Dose-response analysis** is the standard method of testing with lab animals. The response is quantified and graphed (a **dose-response curve**) by measuring the proportion of animals exhibiting negative effects.

  ▪ **$LD_{50}$** values (the lethal dose for 50% of individuals) are calculated from the resulting graph. A high $LD_{50}$ value indicates low toxicity, and a low $LD_{50}$ value indicates high toxicity.

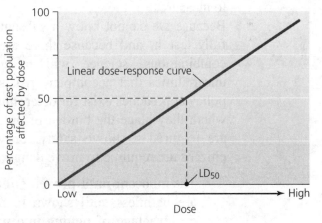

**(a) Linear dose-response curve**

**Figure 14.17a**  Linear Dose-Response Curve

  ▪ **$ED_{50}$ values** (the effective dose for 50%) can be determined when the experimenter is interested in the sublethal effects, including teratogen, carcinogenic, and/or neurotoxic effects.

  ▪ **A threshold dose** is that which elicits a response *only* above a certain dose or threshold.

▌ **Acute exposure** (high exposure for short periods of time) to an environmental toxicant differs from **chronic exposure** (low exposure over long periods of time). Lab tests and $LD_{50}$ values generally reflect acute exposure. Chronic exposure often affects organs gradually, as when smoking causes lung cancer.

▌ **Synergistic effects** are those in which the impact of mixing chemical toxicants becomes greater than the simple sum of their individual effects.

### *How to evaluate risk assessment and risk management, including different philosophical approaches to risk*

▌ **Risk assessment** involves *quantifying* and comparing risks involved in different activities or hazardous substances. This process begins by *identifying* a potential hazard and assessing the risk that *specific concentrations* would have.

▌ Risk can be measured in terms of **probability**, a *quantitative* description of the likelihood of a certain outcome.

▌ Policy decisions made on whether to ban chemicals or restrict their use generally follow years of rigorous lab and environmental testing. Decisions also incorporate economics and ethics and are influenced by political pressure from powerful interests.

▌ Finally, comparing the costs and benefits of addressing the risk(s) is an essential part of **risk management** procedures. However, this can be challenging because the benefits are often economic, whereas the costs often pertain to human or environmental health, which are much harder to measure.

▌ Because we cannot know a substance's toxicity until we measure and fully test it, and because there are so many untested chemicals and combinations, science will never be able to eliminate all the uncertainties that accompany risk assessment. Two philosophical and policy approaches exist for determining safety, and they differ mainly in where they place the burden of proof—whether product manufacturers are required to prove safety, or whether government, scientists, or citizens are required to prove danger.

■ The **innocent-until-proven-guilty approach** assumes that substances are harmless until shown to be harmful. This approach has the disadvantage of putting into wide use some substances that may later turn out to be dangerous. The burden of proof for harm is placed on the public.

■ The **precautionary principle approach** assumes that substances are harmful until shown to be harmless. The burden of proof for safety is on the manufacturer.

■ Many European countries have recently embarked upon a policy course that incorporates the precautionary principle, whereas the United States largely follows the innocent-until-guilty approach.

*How environmental health hazards are regulated in the United States and in the international community.*

▍ In the United States, the Food and Drug Administration (**FDA**) monitors safety of foods, food additives, cosmetics, drugs, and medical devices. The Occupational Safety and Health Administration (**OSHA**, 1970) regulates workplace hazards.

▍ The Environmental Protection Agency (EPA) regulates pesticides under the **Federal Insecticide, Fungicide, and Rodenticide Act (FIFRA)**, 1947.

▍ The EPA regulates all synthetic chemicals not covered by other laws under the **Toxic Substances Control Act (TSCA)**, 1976.

▍ The European Union is currently taking a precautionary principle approach toward testing and regulating manufactured chemicals with its **REACH** (2007) program, shifting the burden of proof for testing chemical safety to the industrial manufacturer.

# Chapter 22: Managing Our Waste

---

## YOU MUST KNOW

- The main types of waste generated.
- The major approaches to managing waste.
- Conventional waste disposal methods: sanitary landfills and incineration.
- The major approaches for reducing waste: source reduction, reuse, recycling, and composting.
- How industrial solid waste is managed.
- The specific issues involved in managing hazardous waste.

---

*The main types of waste generated*

▍ **Waste** refers to any unwanted material or substance that results from a human activity or process.

　▪ **Municipal solid waste (MSW)** is nonliquid waste that comes from homes, institutions, and small businesses, commonly referred to as "trash" or "garbage." It includes everything from paper to food scraps to old appliances and furniture.

　▪ **Industrial solid waste** includes waste from production of consumer goods, mining, agriculture, and petroleum extraction/refining.

- **Hazardous waste** refers to solid or liquid waste that is toxic, chemically reactive, flammable, or corrosive and can include everything from paint and household cleaners to medical waste and industrial solvents.

### The major approaches to managing waste

- Because waste can degrade the water, soil, and air quality, and is also a measure of inefficiency in the industrial process that created the waste, waste management has become a vital human pursuit.
- There are three main components of waste management.

  - **Source reduction**—minimizing waste at its source.
  - **Recycling**—the process of collecting used goods and sending them to facilities that extract and reprocess raw materials that can then be used to manufacture new goods.
  - **Safe waste disposal**—regardless of how effectively we reduce the waste stream through source reduction and recovery, there will likely always be some waste left to dispose. Disposal methods include burying waste in landfills and burning waste in incinerators.

- Paper products comprise the largest component of the municipal solid waste stream in the United States by weight.

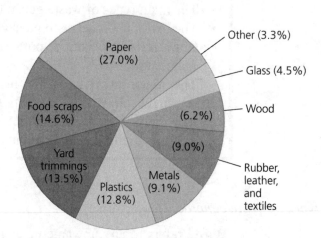

**(a) Before recycling and composting**

**Figure 22.3a**   U.S. MWS Before Recycling and Composting

- The average American generates 2.0kg (4.4 lb) of trash per day, considerably more than people in most other developed nations, primarily because of excess packaging and the use of nondurable goods ("the throwaway society").

### Conventional waste disposal methods: sanitary landfills and incineration

▊ Open dumping and waste burning were the acceptable approaches to dealing with municipal solid waste until human population and consumption rates increased dramatically in the late 20th century.

▊ Since the late 1980s, recovery of materials for recycling expanded significantly. As of 2013, U.S. waste managers were landfilling 53% of MSW, incinerating 13%, and recovering 34% for composting and recycling.

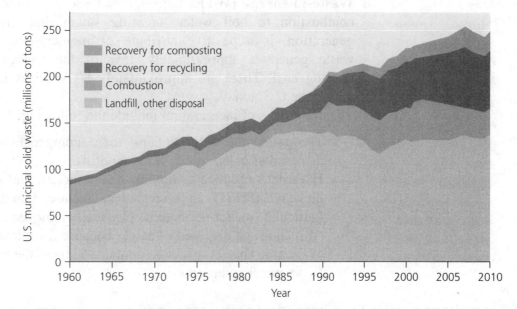

**Figure 22.6** U.S. MSW Management

▊ **Sanitary landfills** are carefully engineered sites where waste is buried in the ground or piled up in large mounds away from wetlands and earthquake-prone sites. Landfills are lined with impermeable surfaces (plastic or clay) to prevent aquifer/groundwater contamination by **leachate** (liquid that results when substances from the trash dissolve in water as it percolates downward). Soil is also layered along with the waste to speed bacterial decomposition, reduce odor, and lessen infestation by pests. Once at capacity, the landfill is closed and capped with an engineered cover that must be maintained as a barrier, preventing water from seeping down and gas from seeping up.

▊ **Landfill gas** (a mix of gases, mostly methane) can be collected, processed and used as an energy resource, in the same manner as natural gas.

▊ The **Resource Conservation and Recovery Act (RCRA)**, enacted in 1976, sets the standards to be met by all sanitary landfills in order to protect against environmental contamination.

▊ The drawbacks of landfills include the following:

▪ The **"not-in-my-backyard" (NIMBY)** reaction causes challenges in finding suitable areas for their location.

- The dry and anaerobic conditions of capped landfills are not conducive environments for bacterial decomposition of the enclosed/buried wastes. Decomposition rates are turning out to be much slower than originally predicted.

- **Incineration**, or combustion, is a controlled process in which mixed garbage is burned at very high temperatures, reducing original weight by up to 75% and its volume by up to 90%.

- **Waste-to-energy (WTE)** facilities that use heat produced by waste combustion to boil water, creating steam that drives electricity generation, is a positive attribute of incineration. When burned, waste generates approximately 35% of the energy generated by burning coal. There are approximately 80 WTE facilities operating across the U.S. today.

- The drawbacks of incineration include the following.

  - The remaining ash contains toxic components that must be disposed of in hazardous waste landfills.
  - Hazardous chemicals, dioxins, heavy metals, and polychlorinated biphenyls (PCBs) can be created and released into the atmosphere.
  - Particulate matter needs to be physically removed from incinerator emissions and disposed of safely because it contains tiny particles of fly ash. This fly ash often contains some of the worst dioxin and heavy metal pollution.

### *The major approaches for reducing waste: source reduction, reuse, recycling, and composting*

- The **3 Rs ("Reduce, Reuse, Recycle")** and composting divert materials from the waste stream.

- **Source reduction** (reducing waste before it is generated) is the *best* waste management approach because it avoids the costs of safe disposal and recycling, helps conserve resources, minimizes pollution, and is economically sound for consumers.

  - Much of the waste stream consists of materials used to package goods and can be reduced with incentives.
  - In 2007, San Francisco became the first U.S. city to ban nonbiodegradable plastic bags.

- **Reuse** is a main strategy to reduce waste: Examples of reusable items include coffee cups and reusable cloth bags.

- **Recycling** has grown significantly in recent years and now removes 25.54% of the U.S. waste stream. U.S. states vary greatly (2–43%) in rates at which citizens recycle.

- The most recent survey of campus sustainability efforts indicates that the average recycling rate was only 29%, leaving definite room for growth. However, in 2016, 350 colleges and universities competed in the annual *Recyclemania* competition, with Richmond College in Dallas, Texas, taking top honors, recycling 82% of its campus waste.

▌ Recycling advocates point out that market prices for recycled goods do not take into consideration external costs (the environmental and health costs of *not* recycling).

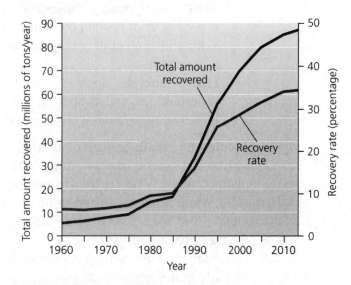

**Figure 22.8**  Recovery Has Risen Sharply in the United States

▌ **Composting** recovers organic wastes and reduces landfill waste, enriches soil, and reduces need for chemical fertilizers.

▌ Financial incentives can help encourage waste reduction and recycling.

   ▪ "Pay-as-you-throw" approach to garbage collection
   ▪ Bottle bills, which encourage consumers to return bottles and cans to stores after use in order to receive a refund

## *How industrial solid waste is managed*

▌ **Industrial solid waste** is defined as solid waste that is considered neither municipal solid waste nor hazardous waste under the U.S. Resource Conservation and Recovery Act (RCRA), and it is managed by state or local governments.

▌ **Industrial ecology** seeks to redesign industrial systems to reduce resource inputs and to maximize both physical and economic efficiency by functioning more like ecological systems and completing a product's **life cycle analysis**.

▌ The Swiss Zero Emissions Research and Initiatives (ZERI) Foundation sponsors dozens of innovative projects worldwide that attempt to create goods and services without generating waste.

## *The specific issues involved in managing hazardous waste*

▌ In most developed nations, hazardous waste generation and disposal is highly regulated. Under RCRA, the EPA sets standards by which states are to manage hazardous waste, and industries are required to track these hazardous materials "from cradle to grave."

■ Two classes of chemicals are particularly hazardous because their toxicity persists over time and they can bioaccumulate: organic compounds and heavy metals.

    ■ **Synthetic organic compounds** and petroleum-derived compounds resist bacterial decomposition; can be readily absorbed through the skin; and can act as mutagens, carcinogens, teratogens, and endocrine disruptors.

    ■ **Heavy metals** such as lead, chromium, mercury, arsenic, cadmium, tin, and copper are fat soluble, break down slowly, and bioaccumulate/biomagnify in food chains.

■ **Electronic waste (e-waste)** is a new and substantially fast-growing source of waste due to today's proliferation of computers, cell phones, and DVDs (currently comprising more than 1% of the U.S. solid waste stream). In terms of disposal, these items contain heavy metals and toxic flame retardants, which easily classify them as a potential hazardous waste.

■ Hazardous waste from industrialized nations is also sometimes dumped illegally in developing nations—a major environmental justice issue.

■ Disposal methods for hazardous waste include the following.

    ■ **Hazardous waste landfills**—require stricter standards and more impervious layers in order to keep hazardous material away from groundwater.

    ■ **Surface impoundments**—shallow depressions lined with impervious material.

    ■ **Deep-well injection**—wells are drilled deep beneath the water table into porous rock, where wastes are injected for long-term storage.

■ The **Comprehensive Environmental Response Compensation and Liability Act (CERCLA)** was established in 1980 as a federal program to clean up U.S. sites polluted with hazardous waste from past activities. The EPA administers this cleanup program, called the **Superfund**. The original objective of CERCLA was to charge the polluting parties for cleanup of their sites, according to the polluter-pays principle; however, for many sites, the responsible parties have never been located.

■ *Love Canal* (New York) and *Times Beach* (Missouri) were the two well-publicized locations of past toxic chemical disposal that eventually led to the passage of CERCLA and the creation of the Superfund.

■ As of 2016, 1,328 Superfund sites remained on the National Priorities List, and only 391 had been cleaned up or otherwise deleted from the list, primarily because of high expense and length of clean-up time.

*Multiple-Choice Questions*

***Use the Figure below to answer questions 1 and 2.***

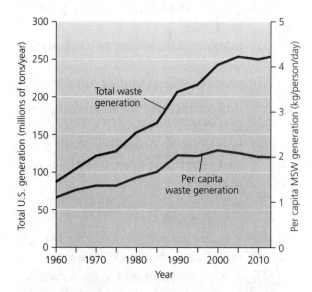

**Figure 22.4** Waste Generation in the U.S. 1960–2010.

1. The per capita waste generation in the U.S. in 2000 was approximately
   (A) 4 kg/person/day.
   (B) 250 kg/person/day.
   (C) 125 kg/person/day.
   (D) 2 kg/person/day.
   (E) 1 kg/person/day.

2. Total waste generation in the U.S. has increased by approximately __% between 1990 and 2010.
   (A) 40%
   (B) 20%
   (C) 10%
   (D) 5%
   (E) 1%

3. Which of the following is *not* an environmental benefit of smart growth?
   (A) impervious surfaces
   (B) decreased fossil fuel consumption
   (C) decreased open space
   (D) decreased water pollution
   (E) increased use of urban growth boundaries

4. An $LD_{50}$ study indicates the
   (A) amount of toxicant it takes to kill half the population.
   (B) lowest dose that kills any member of the population.
   (C) lowest dose that kills 50 out of 1,000 members of the population.
   (D) lethal dose required to kill 5 out of 50 members of the population.
   (E) lowest survival dose of toxicant for 50% of the population.

5. In the United States, approximately how much of our generated municipal solid waste ends up being recovered in composting and recycling?
   (A) about 22%
   (B) about 34%
   (C) > 45%
   (D) >50%
   (E) >75%

6. The federal legislation that imposes a tax on the chemical and petroleum industries to generate funds that pay for the cleanup of hazardous substances is
   (A) the Cradle-to-Grave Act.
   (B) the National Priorities List Act.
   (C) RCRA.
   (D) EPA.
   (E) CERCLA.

7. Which is the *most critical* environmental impact of urban sprawl?
   (A) increased air pollution
   (B) increased acid precipitation
   (C) increased consumption of locally grown produce
   (D) decreased loss of agriculture lands
   (E) decreased reliance on fossil fuels

8. A resource sink is
   (A) a resource that is desirable, is in high demand, and promotes competition between cities.
   (B) an area that produces almost none of the things that it needs.
   (C) an area that is able to trade for all of the resources it needs.
   (D) an area that produces a single resource and is therefore unable to control its price through competition.
   (E) a low area of productivity, especially found in desert areas.

9. Bisphenol A (BPA) is a chemical used in ___ and has the effect of _____.
   (A) herbicides; a neurotoxin
   (B) pesticides; a carcinogen
   (C) plastics; an endocrine disruptor
   (D) synthetic fuels; a carcinogen
   (E) food additives; a teratogen

10. Pesticides in the United States are registered for use through the
    (A) FDA.
    (B) BLM.
    (C) DOE.
    (D) EPA.
    (E) USDA.

11. Currently, which of the following materials comprises the largest component of municipal solid waste by weight?
    (A) e-waste
    (B) heavy metals
    (C) yard waste
    (D) paper
    (E) food scraps

12. Currently, the average American generates approximately ___ lb of municipal solid waste per day.
    (A) 1.2
    (B) 2.2
    (C) 4.4
    (D) 15
    (E) 40

13. The role of urban zoning is to
    (A) provide the necessary revenue to run city governments.
    (B) promote urbanization.
    (C) classify areas for different types of land use.
    (D) integrate agriculture with urban land use.
    (E) promote green building infrastructure.

14. A LEED certification for an urban building indicates that
    (A) it needs to be demolished and new sustainable construction built in its place.
    (B) it meets all the requirements for smart growth.
    (C) it meets the requirements for being built in a sustainable greenway.
    (D) it has been built with consideration for minimum water and energy use.
    (E) it has been built entirely of recyclable materials and is always less expensive than comparable modern buildings.

15. Which of the following are naturally occurring chemicals, potentially toxic to people?
    (A) Bt and *Salmonella*
    (B) carbon dioxide and nitrogen gas
    (C) DDT and DDE
    (D) phthalates and bisphenol A
    (E) crude oil and radon gas

16. Which statement about dose-response studies is *false*?
    (A) Dose-response studies test chemicals across a range of concentrations.
    (B) Dose-response studies test only for lethal effects.
    (C) Dose-response studies can last for days or months.
    (D) $LD_{50}$ values are divided by 10 to determine safe concentrations for wildlife.
    (E) A high $LD_{50}$ value indicates a lower toxicity than that of a chemical with a low $LD_{50}$ value.

17. Which of the following best represents criteria for classifying hazardous waste?
    (A) solid, liquid, gaseous
    (B) municipal, industrial, agricultural
    (C) e-waste, toxic, gaseous
    (D) ignitable, corrosive, reactive
    (E) bioaccumulates, biomagnifies, fat soluble

18. From an environmental waste perspective, which approach is the most effective and desirable?
    (A) Reduce
    (B) Reuse
    (C) Recycle
    (D) Compost
    (E) Incinerate

***Use the Figure below to answer questions 19 and 20:***

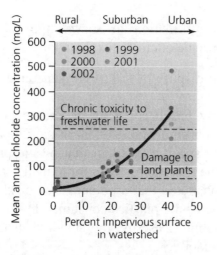

19. The important conclusion the data in the graph shows is that the salt concentration
    (A) increases with time.
    (B) increases as asphalt pavement coverage increases.
    (C) decreases with time.
    (D) decreases as asphalt coverage increases.
    (E) in suburbia regularly produces chronic toxicity to freshwater life.

20. The data indicate that with a 35% impervious surface, one can expect
    (A) chronic toxicity to freshwater life.
    (B) 400 mg/L chloride.
    (C) 50 mg/L chloride.
    (D) damage to land plants.
    (E) cannot be determined from available data.

21. Which of the following environmental hazards is most capable of biomagnification?
    (A) radon and $Pu^{239}$
    (B) lead
    (C) asbestos
    (D) PCB
    (E) PAHs found in cigarette smoke

22. Many chemical substances are very challenging for risk assessment studies because of their
    (A) hemorrhagic effects.
    (B) persistence in the environment.
    (C) quick disappearance and biological decay in the environment.
    (D) allergic effects, making them unsuitable for human studies.
    (E) synergistic effects.

23. E-waste is a major source of
    (A) heavy metals.
    (B) radioactive materials.
    (C) combustible inorganic compounds.
    (D) ignitables.
    (E) acid corrosives.

24. Incineration is a contemporary method of handling municipal solid waste and contributes to all of the following except
    (A) generating electricity.
    (B) increasing air pollution.
    (C) reducing volume.
    (D) extracting energy.
    (E) decreasing the life of sanitary landfills.

*Free-Response Question*

The Jones family lives in urban Boston, MA, in a 100-year-old historic house. This family of four is planning on adding more sustainable approaches to their daily living. Initially, they plan to focus on their municipal waste stream (MSW), followed by reducing their daily exposure to environmental toxins. Currently, they generate 1.8 kg (4 lb) of MSW per day per person. The disposal fee in their city is $100 per ton (2,000 lb).

The table below describes the percentage composition of the typical municipal solid waste (MSW) generated by the Jones family.

| Material | Percentage (%) |
|---|---|
| Yard waste/grass clippings | 13 |
| Food scraps | 13 |
| Paper and cardboard | 33 |
| Plastic | 12 |
| Metals | 8 |
| Glass | 5 |

Use the preceding table and information to answer the following questions. Show all work and provide complete explanations.

a) What is the annual cost to dispose of the MSW generated by the Jones household?

b) Recently, the Jones family has become aware of the potential problem of plastic as an environmental toxicant in their municipal waste stream. Explain one specific problem that plastic presents for the environment and one specific problem for human health.

c) Recent research has demonstrated that a potential toxic substance "X" has been found in one of the plastics consumed by the Jones family and the general public. Describe the standard method of determining its potential toxicity.

d) The Jones family would like to receive a *LEED certification* for their historic home; however, they realize that they will have to make several changes. Describe two specific changes along with the reason for making the change that will allow them to receive a LEED certificate.

# Water Resources and Pollution

## *Chapter 15: Freshwater Systems and Resources*

---

**YOU MUST KNOW**

- Water's importance and the distribution of freshwater resources on Earth.
- The major types of freshwater systems.
- The various categories of water use by humans and the problems associated with the current utilization of this resource.
- The solutions that can effectively address freshwater depletion issues.
- The main categories of water pollutants and how they affect water quality.
- The main procedures involved in drinking water and human wastewater treatment.

---

### *Water's importance and the distribution of freshwater resources on Earth*

- 97.5% of Earth's water resides in the oceans. Only 2.5% of Earth's water is considered fresh water, with most of this scarce resource tied up in glaciers, icecaps, and underground aquifers.
- Only 1% of fresh water is easily accessible on Earth's surface. Water is unevenly distributed across the planet's surface, from 470 inches per year (on the tropical Hawaiian island of Kauai) to virtually zero (in Chile's Atacama Desert).
- Humans are *not* distributed across the globe in accordance with water availability. Moreover, there is a growing number of heavily populated countries located in arid lands where fresh water is scarce.

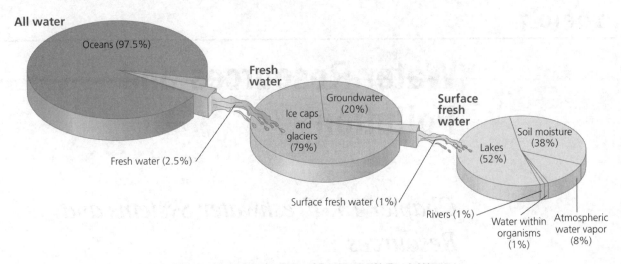

**Figure 15.3** Distribution of Earth's 2.5% Fresh Water

- Water is essential and maintains all life and ecosystems. It regulates the temperature of the planet, redistributing heat, and cycles essential nutrients through land, air, and all living things.
- Fresh water is a renewable resource because of the hydrologic cycle, which returns water to both surface and groundwater sources from the atmosphere. However, *only if* water is used and managed sustainably can water be realized as a renewable resource. Currently, in many industrial nations, concerns over water *quantity* dominate, whereas in developing countries, concerns over both water *quantity* and *quality* persist.
- California's ongoing drought is an example of the water shortages faced by many parts of the world today. California has experienced record-low levels of precipitation combined with record-high heat since 2012, with devastating effects on its agricultural sector, including a reduction of hydroelectric power capacity, and extensive damage to its forests, with more than 66 million tree deaths.
- It is expected that global climate change will further worsen the disparity between water availability and human needs by altering precipitation patterns, melting glaciers, and intensifying droughts and flooding, producing a great need for sustainable, long-term water conservation programs.

### The major types of freshwater systems

- **Surface water** is the fresh water that exists above ground including streams, rivers, ponds, lakes, and wetlands. The area of land drained by a river system is referred to as a **watershed**, or drainage basin. The Mississippi watershed is the largest in the U.S., draining over 40% of the land area of the lower 48 states.

- Nearly three-quarters of all water used in the United States comes from surface water sources and is used for electric power generation, municipalities, industrial use, and irrigation.
- Bodies of water are often classified according to their productivity. **Oligotrophic** lakes and ponds have low nutrients and high oxygen levels; **eutrophic** water bodies are high in nutrients and low in oxygen. Changes in a body of water can occur naturally; however, eutrophication now usually results from human-caused excessive nutrient pollution.
- Excessive surface water withdrawals can drain rivers and lakes. The Colorado River is a prime example, as it often fails to reach the Gulf of Mexico because of over extraction by both Mexican and U.S. farmers.

▌ **Wetlands** are surface water systems in which the soil is saturated with water and which generally feature shallow standing water with noticeable vegetation. Marshes, swamps, and bogs are types of wetlands, all of which are enormously rich and productive. **Vernal pools** are seasonal wetlands containing water only at a specific time of the year (springtime, in the Northern Hemisphere).

- Wetlands are extremely valuable as habitat for wildlife, providing important ecosystem services by slowing runoff, reducing flooding, recharging aquifers, and filtering pollutants.
- Extensive drainage of wetlands for agricultural and urban development has occurred throughout the United States; less than half of all wetlands remain.
- Wetlands can also be lost when water is withdrawn or diverted, rivers channelized, and dams are built. A prime example of extensive wetlands loss includes the Louisiana delta region near New Orleans.
- *The Ramsar Convention on Wetlands of International Importance* (1971) seeks to protect the many ecological, social, and economic benefits of wetlands worldwide within the context of sustainable development.

▌ **Groundwater** is water beneath Earth's surface held within the pores of soil and/or rock. Although groundwater sources account for only one-quarter of all water used in the United States, they supply almost half of all drinking water and are an important source of irrigation for Midwest cropland.

- An **aquifer** is an underground reservoir contained in porous, spongelike formations of rock, sand, or gravel. The largest known aquifer is the Ogallala Aquifer, located under eight U.S. Great Plains states.

- As aquifers are withdrawn, they recharge *very* slowly; average rates of groundwater flow are only about 1 m per day. Therefore, groundwater is often more easily depleted than surface water because of this slow recharge.
- **Water tables** (the upper limit of groundwater held in an aquifer) drop when aquifers are overdrawn; as a result, groundwater becomes more difficult and expensive to extract. In parts of Mexico, India, China, and many Asian and Middle Eastern nations, water tables are falling 3–10 ft per year.

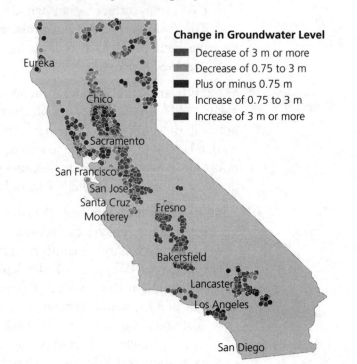

**Figure 15.14** Change to California Groundwater Levels during Past Decade

- In coastal areas, where extraction of groundwater is high, **saltwater intrusion** can affect an aquifer, making water undrinkable, with land subsiding and the creation of **sinkholes**.

## *The various categories of water use by humans and the problems associated with the current utilization of this resource*

- **Agriculture** accounts for approximately 70% of Earth's annual fresh water use.
  - The amount of land currently being irrigated has doubled in the last 50 years to meet the increased demand for human food.
  - Most irrigation remains highly inefficient, and crops end up using only 40% of the water applied by "flood and furrow" irrigation methods.
  - Overirrigation leads to the **salinization** of agricultural lands, reducing crop production.

- Worldwide, about 15–35% of water withdrawals for irrigation are thought to be unsustainable. Often, the Colorado River and Rio Grande no longer flow to the Gulf of California as a result of the extensive diversions and current consumption rates these rivers have undergone.

▌ **Industry** accounts for approximately 20% of Earth's annual fresh water use.

▌ **Residential use** accounts for about 10% of Earth's fresh water use.

- Bottled drinking water production now is a significant groundwater extraction and depletion problem, representing a $14 billion bottled water industry.
- Americans now drink more bottled water than beer or milk, and pay more per gallon for it than for gasoline.
- Bottled water exerts substantial *ecological* impact because of heavy packaging, transportation over long distances using fossil fuels, and, in the United States, failure to recycle roughly 3 of 4 plastic water bottles.
  - Major U.S. cities including New York City, San Francisco, and Seattle now prohibit using government funds to purchase bottled water. As of 2015, over 90 colleges and universities in the U.S. had banned the sale of plastic water bottles.
- Bottled water is not demonstrably safer or healthier than tap water. In fact, one often does not know its source, unlike municipal tap water, which is strictly regulated by the U.S. Environmental Protection Agency (EPA).

▌ **Water diversion projects** have long been a part of human history, bringing surface water to farm fields, homes, and cities through the use of dams, aqueducts, pipes, and open-air canals.

- California has one of the most extensive freshwater diversion schemes in the world. The Colorado River is a prime example of the use of dams (13 major dams store water in enormous reservoirs), open-air canals, and aqueducts (approximately 1,200 miles),which have altered the once wild Colorado River to divert water. The result is that the diverted water is used mostly to irrigate crops in desert regions and to sustain the growth of Los Angeles, San Diego, Phoenix, Tucson, and Las Vegas.
- California's arid Imperial Valley is the nation's largest irrigation district. The source of this federally subsidized water is the extensively diverted Colorado River.
- The world's largest dam project is the Three Gorges Dam on China's Yangtze River, completed in 2008 at a cost of $39 billion. Its reservoir flooded 22 cities, requiring the largest human resettlement project in China's history.

■ **Channelization** (straightening, deepening, widening, or lining existing stream channels) is an engineering technique that has been extensively used to transport water, as well as for flood control and drainage improvement. This process has caused extensive habitat destruction and fragmentation in many parts of the world.

▮ To protect urban and residential areas against floods, governments have built **dikes** and **levees** (long raised mounds of earth) along riverbanks to hold water in the main channels. However, in the long term and in rural agricultural areas, floods are immensely beneficial because floodwaters build and enrich flood-plain soil by spreading nutrient-rich sediments over large areas. Overall, human development tends to worsen flooding because pavement and compacted soils speed runoff, sending intense pulses of water into rivers during storm events.

▮ The diverse costs and benefits of damming rivers and streams are currently being reevaluated in the United States. The largest U.S. dams are on the Colorado River (Hoover Dam and Glen Canyon Dam), each holding back a reservoir (Lake Mead and Lake Powell), which can be utilized for hydroelectric energy generation. Habitat alteration is one of the main negative impacts of a dam's construction and presence.

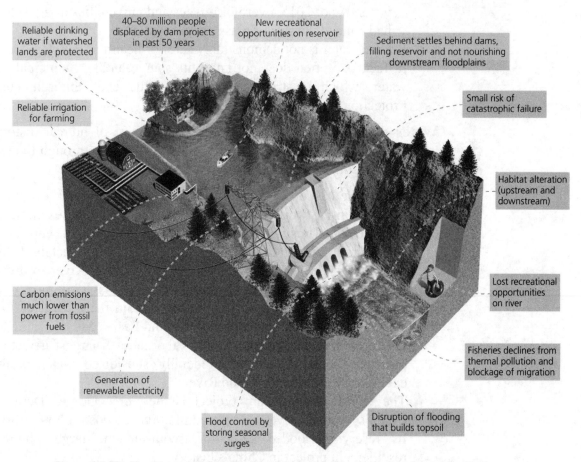

**Figure 15.18** The Diverse Consequences of Damming Rivers for People and the Environment

- China's Three Gorges Dam reservoir is depriving the river downstream of valuable sediment, eroding tidal marshes at the Yangtze's mouth, and degrading Shanghai's coastal environment. Recently, the potential for earthquake damage to the dam, perhaps even leading to its collapse, has been discussed.

- Some dams are being removed in areas where it appears, in retrospect, that the costs outweigh the benefits. By removing a dam, riparian ecosystems can be restored, economically valuable fisheries can be reestablished, and river recreation such as fly-fishing and rafting can be increased. The Edwards Dam on Maine's Kennebec River was one of the first to be removed (1999) in the United States. In 2014, the world's largest dam removal project was completed for the Glines Canyon Dam in Washington state.

- Political tensions over water may heighten in the future as depletion of fresh water continues to lead to shortages and resource scarcity.

  - Many predict that water's role in regional conflicts will increase as human population continues to grow and as climate change continues to alter precipitation patterns.

  - "Water promises to be to the 21st century what oil was to the 20th century: the precious commodity that determines the wealth of nations." (*Fortune* magazine, May 2000).

  - **Water mining**, withdrawing water faster than it can be replenished, is now taking place in California's central agricultural regions, an area of important food production for the United States. In parts of Mexico, India, China, and many Asian and Middle Eastern nations, water tables are also falling 1–3 meters per year.

### *The solutions that can effectively address freshwater depletion issues*

- The most effective solutions to water depletion issues involve *reducing demand*; that is, focusing on *conserving* water and *increasing efficiency* of water use.

  - For *agriculture*: **efficient irrigation methods** (drip irrigation and low-pressure spray irrigation) and lining irrigation canals to prevent leaks all result in a dramatic decrease of water lost to evaporation. Drip irrigation improves efficiency by delivering smaller volumes of water directly to crop roots for absorption.

  - Choosing **crops to match the land and climate** in which they are farmed can save huge amounts of water.

  - **Selective breeding and genetic modification** can produce crop varieties that require less water.

  - **Recycling treated municipal wastewater** for irrigation purposes conserves water.

- For the *residential* sector: households can reduce water use by installing **low-flow faucets, showerheads, toilets,** and **washing machines**.
  - Repairing leaks in pipes and water mains conserves a considerable amount of water.
  - Automatic dishwashers use less water than washing dishes by hand.
  - Using **gray water** (the wastewater from showers and sinks) to water lawns and **xeriscaping** (landscaping using plants adapted to arid conditions) are effective in reducing water demand.

- Attempts to *increase the supply* (through temporarily increasing extraction or new diversion projects) may solve water shortages in one area, but they often cause new shortages in distant areas and have not proved to be sustainable in the long term. Desalination approaches are increasing supplies in some locations, with over 20,000 facilities operating worldwide.

  - **Desalination**, or desalinization, is the removal of salt from seawater or other water of marginal quality. Although increasing a local fresh water supply, desalination is expensive, requires large inputs of fossil fuel, kills aquatic life at water intakes, and generates a very concentrated salty waste.
  - **Reverse osmosis** (microfiltration) and **distillation** are the two methods used in desalination.
  - In late 2015, a San Diego County desalination plant, the largest in the U.S., began supplying this southern California region with 190 million L/day of freshwater.

- **Market-based approaches** to water conservation are being debated. Will ending government subsidies of inefficient practices and allowing water to become a commodity (so that the price reflects the true costs of extraction) become solutions of the future? Market-based approaches support the idea that waste, inefficiency, and environmental degradation often result when prices do not reflect the scarcity of water.

- The **privatization of water** supplies, especially in developing countries, has also become an area of debate. In a private system, individuals buy, sell, and trade water rights rather than governments controlling all access to water. However, critics charge that private water markets will undersupply consumers, leading to inequitable distribution because private firms often have little incentive to allow equal access to water for rich and poor alike.

### The main categories of water pollutants and how they affect water quality

- **Water pollution** is the release into the environment of matter or energy that causes undesirable impacts on the health or well-being of humans or other organisms. In 2013, the EPA reported that 55% of the 2,000 U.S. streams and rivers sampled were in a too degraded condition to support aquatic life.

▌ **Point sources** of water pollution come from discrete locations or specific spots, such as a factory or sewer pipe. In contrast, **non-point sources (NPS)** of pollution are cumulative, coming from multiple inputs over larger areas, such as runoff from farms, city streets, or residential neighborhoods. The U.S. Clean Water Act (1972, 1977) has addressed point-source pollution with some success; however, non-point source pollution continues to be more problematic to control from such common activities as applying fertilizers and pesticides to lawns and applying salt to roads in winter.

▌ The main categories of water pollution are:

1) **Toxic chemicals:**
   ♦ *Organic chemicals:* Pesticides (e.g., DDT), herbicides (e.g., atrazine), and petroleum products (e.g., oil) can directly poison plants and animals; some have been known to be human carcinogens. Some have produced a wide variety of adverse effects in the natural world. For example, DDT has caused thinning of bald eagle and brown pelican eggshells. Other adverse effects have resulted from ocean oil spills including the 1989 *Exxon Valdez* spill in Alaska's Prince William Sound and the 2010 British Petroleum's *Deepwater Horizon* oil contamination of the Gulf of Mexico.
   ♦ *Heavy metals:* Lead, mercury, and arsenic most often enter the water supply from industrial discharges and some household chemicals. These have been linked with cancer and disruption of the immune/endocrine systems.
   ♦ *Inorganic chemicals:* Acids, bases, and salts can enter the water supply as a result of the activities of industry, households, and mining. Acid precipitation damage to aquatic ecosystems (acid rain) has been highly documented and acid mine drainage (water with a high concentration of sulfuric acid) from coal and metal mines causes significant water pollution.
   ♦ Leakage of these toxic chemicals from underground oil and gasoline storage tanks is a current focus of major EPA cleanup efforts across the United States.

2) **Sediments:** When poor farming or forestry practices cause large-scale disturbance of vegetation, large sediment loads may damage aquatic systems by blocking out sunlight, impairing the normal photosynthetic processes and food chain relationships.

3) **Thermal pollution:** Water is often used to cool many industrial processes, and when this warm water is returned to the environment its ability to hold oxygen is greatly reduced. Cold water can carry more dissolved oxygen than warm water and aquatic life (fish and aquatic invertebrates) are very sensitive to changes that lower oxygen levels.

4) **Oxygen-demanding wastes:** Organic matter (human sewage or animal wastes) nourishes the growth of microbes that are decomposers, and as the decomposition process proceeds, oxygen is consumed. This is referred to as **biochemical oxygen demand**, or **BOD** (the amount of oxygen required to break down organic matter over a specific period of time). High BOD values indicate that a water body is more polluted by organic wastes than one with a lower BOD value.

5) **Nutrient pollution:** Nitrates and phosphates contained in agricultural runoff of synthetic fertilizers from Midwest farms annually leads to eutrophication and hypoxia in both coastal marine waters (Gulf of Mexico's "dead zone") and inland lakes and ponds. Research has indicated that in order to alleviate these "dead zones," it will be necessary to further reduce phosphorus pollution from industry and sewage treatment, as although phytoplankton needs both N and P, phosphorus has become the primary limiting factor.

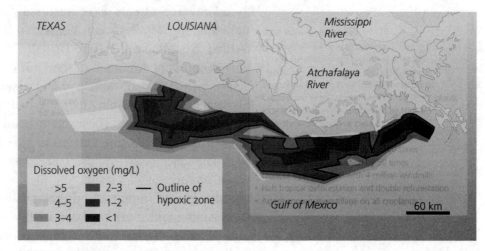

**(a) Dissolved oxygen at ocean bottom**

**Figure 1a**   Dissolved Oxygen Concentration in Gulf of Mexico's Bottom Water off the Louisiana Coast in 2015 (Regions considered hypoxic [<2 mg/L] are encircled with a black line)

6) **Pathogens and waterborne diseases:** Disease-causing organisms can enter any body of water that is contaminated with human waste from inadequately treated sewage or with animal wastes from feedlot run-off.

   ♦ Pathogenic viruses, protists, and bacteria cause more human health problems than any other type of water pollution.

   ♦ Drinking water treatment facilities are designed to reduce these pathogenic risks by treatment using chemicals and other disinfectant (UV) processes.

   ♦ Scientists who monitor water quality use a variety of biological, chemical, and physical indicators/tests.

- **Biological indicators** include testing for the presence or absence of **fecal coliform** bacteria, which is the criterion used by the EPA to establish safe drinking water standards (0 fecal coliform/100 ml sample). **Aquatic invertebrate biodiversity** measurements are also used as biological indicators of water quality and ecosystem/ stream health.
- **Chemical testing** includes determining pH, nitrate and phosphate content, and, most important, dissolved oxygen content.
- **Physical properties** of water that can reveal water quality are color, turbidity, and water temperature.

### *The main procedures involved in drinking water and human wastewater treatment*

- Drinking water treatment is widespread in developed nations today; however, in developing countries the lack of safe drinking water is often the most serious water quality issue, and the one causing the most deaths. Many developing countries also lack any type of formal sewage treatment, especially in rural areas. Urban wastewater treatment in most developed countries relies on primary and secondary treatment of sewage.
- **Drinking water treatment** consists of the following.

  - Reservoir or aquifer water is pumped to the facility where it is first treated with chemicals to remove any particulate matter.
  - Water is then passed through filters of sand, gravel, and charcoal to further remove impurities.
  - Finally, the water is treated with chlorine and/or undergoes UV treatment for its final disinfection.

- **Wastewater** refers to water that people have used in some way. It includes water carrying sewage; water from showers, sinks, washing machines, and dishwashers; water used in manufacturing or industrial cleaning processes; and storm water runoff.
- **Septic systems** are methods of wastewater disposal in rural areas. The septic tank is designed to separate solids from liquid and to digest (biochemically change) and store organic matter for a while, allowing the clarified liquid to discharge into an underground drain/leach field and into the surrounding soil. The combination of pipes and soil makes up the leach field, where the decomposing septage is further degraded by soil microorganisms.

◗ **Municipal wastewater treatment** is routine in urban, densely populated areas in developed countries. Primary and secondary treatment are required by U.S. federal law and consists of the following.

   ◗ **Primary treatment:** the physical removal of contaminants in settling tanks or clarifiers. Approximately 60% of suspended solids are removed.

   ◗ **Secondary treatment:** a biological process that involves the wastewater proceeding to aeration basins where it is stirred and oxygenated so that aerobic bacteria can degrade the organic pollutants. Roughly 90% of the suspended organic solids are removed.

   ◗ **Tertiary or advanced treatment:** a physical and chemical process removing specific pollutants left in the water after primary and secondary treatment. For example, this process could remove a specific toxic chemical (heavy metal) resulting from a local industrial plant, or excess nitrates from water contaminated with excess fertilizers. This is an expensive process and exists in only a few urban locations.

   ◗ Finally, the water is treated with chlorine, ozone, or ultraviolet light to kill infectious agents, before the effluent is piped into the receiving river or the ocean.

   ◗ **Sludge** is the solid material that is removed during the wastewater treatment process. It is partially dried on site and then either taken to the local municipal incinerator or disposed of in a landfill. Occasionally, some sludge is used as fertilizer on cropland in local areas where toxic industrial and manufacturing byproducts are not present in wastewater.

◗ In summary, *preventing* water pollution is easier, much more economic, and more effective than mitigating it later. Water quality issues continue to take a toll on the health, economies, and societies of both developed and developing nations. Water depletion issues are becoming a critical concern in many areas of the developing world, as well as in arid regions of all developed nations.

# Chapter 16: Marine and Coastal Systems and Resources

---

**YOU MUST KNOW**

- The ocean's role in climate regulation and how this is accomplished.
- The major types of marine ecosystems.
- The various categories and effects of the major marine pollutants.
- The current state of marine fisheries and the reasons for their decline.
- How to evaluate the effectiveness of marine conservation efforts as solutions to the major marine environmental concerns.

---

### *The ocean's role in climate regulation and how this is accomplished*

▍ The world's oceans cover 71% of Earth's surface and contain 97.5% of its water.

▍ By absorbing heat and releasing it to the atmosphere, the oceans help regulate Earth's climate. They also influence climate by moving heat from place to place by surface currents flowing horizontally (such as the Gulf Stream) from equatorial regions to Europe, moderating these higher latitudes.

▍ Surface winds and heating also create vertical currents. **Upwellings**, the upward flow of cold, deep water rich in nutrients toward the surface, create biologically rich regions (upwelling zones off Peru's coastline), producing highly productive fisheries.

▍ **Thermohaline circulation** is a worldwide current system in which warmer, fresher water moves along the surface and colder, saltier water (which is more dense) moves deep beneath the ocean's surface. Scientists hypothesize that interrupting the thermohaline circulation could trigger rapid climate change. For example, if global warming causes much of Greenland's ice sheet to melt, the resulting freshwater runoff into the Atlantic Ocean could stop the **North Atlantic Deep Water (NADW)** formation, shutting down the northward flow of warm water.

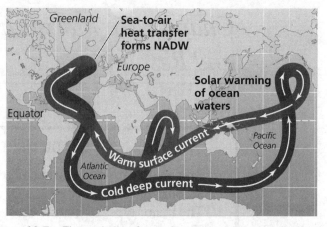

**Figure 16.7** Thermohaline Ocean Currents and the North Atlantic Deep Water (NADW) Circulation

▮ **El Niño-Southern Oscillation (ENSO)** is the exceptionally strong warming of the eastern Pacific Ocean that occurs every 2 to 7 years, depressing local fish and bird populations and significantly altering the marine food web. It is caused by a systematic shift in atmospheric pressure (from high to low), causing warm surface water from the western Pacific to flow eastward. This suppresses upwelling along the Pacific coast of the Americas, shutting down the delivery of nutrients that support marine life. La Niña events are the opposite of El Niño events.

▮ Since the Industrial Revolution, the oceans have sequestered roughly a third of the excess $CO_2$ that we've added to the atmosphere (through burning fossil fuels and forest clearing), slowing the onset of global climate change; however, the oceans may soon be reaching saturation, with climate change accelerating.

    ■ **Ocean acidification** is the process by which today's ocean is becoming more acidic (lowering the pH) as a result of increased carbon dioxide moving from the atmosphere into the oceans. When ocean water absorbs $CO_2$ from the air and forms carbonic acid and bicarbonate ions, the calcium carbonate shells of marine creatures, such as corals, begin dissolving.

    ■ The chemical reactions involved in **ocean acidification** begin with the excess carbon dioxide (now found in the atmosphere) reacting with water and forming carbonic acid: $CO_2 + H_2O \rightarrow H_2CO_3$

    ■ This carbonic acid dissociates into bicarbonate ions and hydrogen ions: $H_2CO_3 \rightarrow HCO_3 . + H^+$

    ■ The resulting hydrogen ions then combine with carbonate ions ($CO_3^{2-}$) to form additional bicarbonate ions: $H^+ + CO_3^{2-} \rightarrow HCO_3 .$

    ■ As the availability of carbonate ions in ocean water declines, it becomes more difficult to build calcium carbonate shells (the process known as calcification) by marine organisms. In fact, as carbonic acid and bicarbonate ions become more abundant in ocean water, the calcium carbonate shells of marine creatures begin dissolving.

- Because coral reefs provide billions of dollars' worth of ecosystem services, the potential decline of coral reefs due to ocean acidification now considered to be one of the more serious consequences of global climate change.

### The major types of marine ecosystems

- The **open ocean**, far from the influence of land ecosystems, varies in its biodiversity and productivity, with most of this environment having significantly lower primary productivity than the many types of coastal marine ecosystems found around its vast perimeter.

  - **Coastal/intertidal** areas are often some of the most productive as a result of upwellings and/or their daily nutrient exchange with land. **Salt marshes** line temperate shorelines, have extremely high primary productivity, and provide critical habitat for shorebirds and many commercially important fish and shellfish. They also filter pollution and stabilize shorelines against storm surges.
  - **Mangroves** (salt-tolerant trees) line coasts in the tropics and sub-tropics, and with their partially aerial roots, provide essential habitat and nurseries for fish, shellfish, and many invertebrate species.
  - **Estuaries** are bodies of water where rivers flow into the ocean, mixing fresh water with saltwater, and producing biologically productive ecosystems. Globally, estuaries have been affected by coastal development, water pollution, habitat alteration, and overfishing. Chesapeake Bay, San Francisco Bay, and Washington's Puget Sound are all estuaries that have experienced significant development and human impact.
  - **Coral reefs** are composed of the calcium carbonate skeletons of tiny marine animals (corals and their symbiotic zooxanthellae) living in shallow subtropical and tropical waters. They are experiencing alarming declines worldwide due to "coral bleaching" (thought to be caused by increased sea surface temperatures associated with global climate change), nutrient pollution (which promotes algal growth, smothering the corals), and ocean acidification.
  - **Kelp forests** consisting of large brown algae, or kelp, grow along many temperate coasts, supplying shelter and food for many invertebrates and other marine organisms (the keystone species, sea otters, live in kelp forests). Kelp forests also absorb wave energy and protect shorelines from erosion.

### The various categories and effects of the major marine pollutants

- The oceans have been a sink for human sewage, municipal trash, and industrial wastes for centuries as a result of the misconception that oceans have an infinite capacity to absorb and dilute these toxic pollutants. The major marine pollutants are as follows.

▌ **Oil** is often the first category of marine pollutant that comes to mind because about 30% of our crude oil and nearly half of our natural gas comes from seafloor deposits in regions such as the North Sea and Gulf of Mexico. Of the 1.3 million metric tons of petroleum entering the world's oceans each year, nearly 50% is from natural marine seeps and 38% from non-point sources originating on land. The majority does NOT come from tanker spills or offshore platform failures.

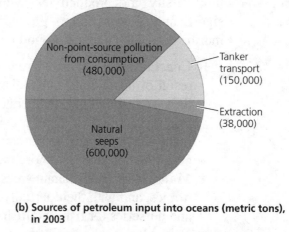

**(b) Sources of petroleum input into oceans (metric tons), in 2003**

**Figure 16.16**   Oil Spills from Tankers on the Decline

▇ Fishing industries and tourism industries are economically devastated when oil is spilled, as in the 2010 British Petroleum's Deepwater Horizon offshore drilling platform explosion in the Gulf of Mexico.

▇ The **U.S. Oil Pollution Act of 1990** created a $1 billion oil pollution prevention and cleanup fund. It also requires that all oil tankers in U.S. waters be equipped with double hulls by 2015.

▇ Sea turtles, dolphins, and birds die when they swallow oil, when oil interferes with breathing, or when oil coats birds' feathers, leading to hypothermia.

▌ **Plastic debris** and **discarded fishing nets** that are lost or intentionally discarded can snare and drown both fish and marine mammals. The North Pacific Gyre, the region ringed by the clockwise flow of currents in the northern Pacific Ocean, is now often referred to as "the **Great Pacific Garbage Patch**" because of the density of plastic trash in the water.

▌ Computer modeling is currently following plastic and other marine ocean debris and was able to predict the 2014 arrival of the Tohoku earthquake/tsunami (2011) debris on the western coast of the U.S.

▌ **Mercury** and other toxic pollutants can make fish and shellfish unsafe for human consumption. Mercury, a toxic heavy metal emitted from coal combustion, mine tailings, and other sources, settles on land and water, and *bioaccumulates* and *biomagnifies* in animals' tissues. As a result, fish (swordfish and tuna) and shellfish at high trophic levels can contain substantial levels of mercury, making them particularly dangerous for children and pregnant women to consume.

▌ **Excess nutrients** from fertilizer runoff or other nutrient inputs can create dead zones (hypoxic conditions) in coastal marine ecosystems (e.g., the Gulf of Mexico's dead zone extending from the mouth of the Mississippi River). In addition, sometimes these excessive nutrient concentrations produce population explosions of harmful algal blooms that produce powerful neurotoxins that affect the vertebrate nervous systems. Red tides are often the name given to these toxic algal blooms that cause death among zooplankton, birds, fish, marine mammals, and humans.

### *The current state of marine fisheries and the reasons for their decline*

▌ Over 50% of the global marine fish populations are fully exploited and cannot be harvested further without depleting them. An additional 28% of marine fish are overexploited and are already being driven toward extinction. Only about 20% of global marine fish populations can actually yield more than they are already yielding without being driven into decline.

▌ Total global fisheries catch, after decades of increases, leveled off after about 1988 despite increased fishing effort. The U.N. Food and Agriculture Organization (FAO) then concluded that "the maximum wild capture fisheries potential from the worlds' oceans has probably been reached."

▌ The collapse of the major cod fisheries on Canada's Grand Banks (off Newfoundland) and Georges Bank (off Massachusetts) resulted from overfishing and bottom-trawling (fishing by dragging weighted nets across the seafloor, destroying underwater habitat).

■ By 1992, scientists reported that mature cod were at just 10% of their long-term abundance, and the Canadian Fisheries announced a two-year ban on commercial cod fishing off Labrador and Newfoundland. When cod stocks did not rebound by 1994, the moratorium was extended, and in 2003 the cod fisheries were closed indefinitely to recreational as well as all commercial fishing.

■ In 2009, a portion of the Great Banks off the southeastern coast of Newfoundland was reopened after data showed that the cod stocks were recovering slightly.

■ In U.S. waters, cod stocks also collapsed, and in 1994 the National Marine Fisheries Service (NMFS) closed the prime fishing areas on Georges Bank. In 2008, managers announced that these stocks were only 12% of what they needed to be for sustainability, and today scientists continue to struggle to explain why cod are not recovering.

■ In the absence of trawling on Georges Bank for over a decade, sea-floor invertebrates like scallops have increased in biomass 14-fold.

■ Today's **industrialized commercial fishing** fleets employ huge factory vessels using fossil fuels and use several methods to capture fish that are highly efficient but also environmentally damaging. Most of the current commercial fishing practices also capture nontarget organisms, referred to as **bycatch**, resulting in the deaths of millions of fish, sharks, marine mammals, and birds each year. A 2011 study from NOAA reports that 17% of all commercially-harvested fish are captured unintentionally. The industrialized commercial fishing methods include the following:

   ■ **Driftnets**—Miles-long transparent nylon nets are set out to drift just below the surface in open ocean water to capture schooling fish such as bluefin tuna, swordfish, and some squid species. This method has often been referred to as "strip-mining" the oceans because it is so very destructive in the removal of *all* marine life from large expanses of the worlds' oceans. Driftnet fishing was banned by the United Nations in 1992 and the European Union in 2002 because of the large bycatch problem.

   ■ **Long-lining**—Extremely long lines (up to 50 miles) with numerous baited hooks are set out in open ocean water. Tuna and swordfish are among the species targeted.

   ■ **Bottom-trawling**—Weighted nets are dragged along the floor of the continental shelf to catch groundfish (cod) and other benthic organisms such as scallops. Bycatch is a problem, as is benthic habitat destruction for many marine species.

■ In just a decade after the arrival of industrialized fishing, catch rates often drop precipitously, with 90% of large-bodied fish and sharks eliminated. This means that the oceans today may contain only one-tenth of the fish they once did. Also, when animals at high trophic levels are removed from a food web, the proliferation of their prey can alter the nature of the entire community, and many scientists conclude that most marine environments may have been very different prior to the current industrial fishing practices.

■ Some of these environmental effects may be partially masked because fishing fleets are traveling longer distances, fishing in deeper waters, spending more time fishing, and setting out more nets and lines. However, as fishing pressures have increased, the size and age of fish caught have declined, and fleets have also been targeting fish species less desirable (changing from the popular cod to the much smaller capelin), a phenomenon referred to as "fishing down the food chain."

■ When fish biodiversity is reduced because of overfishing, many marine ecosystem services have been found to be threatened. The reduction in marine nurseries and reduced filtering and detoxification (provided by wetland vegetation and oyster beds) often lead to harmful algal blooms, dead zones, fish kills, and beach closures.

*How to evaluate the effectiveness of marine conservation efforts as solutions to the major marine environmental concerns*

▌ **Maximum sustainable yield** (harvesting no more than half of any given population) approaches taken by fishery managers are not always working, as many fish stocks are continuing to plummet. Many now feel that **ecosystem-based management**, shifting the focus away from individual species and toward viewing marine resources as elements of larger, connected ecological systems, is the more appropriate approach.

▌ **Marine protected areas (MPAs)** do not always protect marine species, as most allow commercial fishing and other extractive activities. Only 3% of the world's ocean areas are currently designated MPAs.

▌ **Marine reserves** are protected areas that are "no take" in terms of harvesting marine species. These are designed to preserve ecosystems intact and to improve fisheries for the long-term. Unlike terrestrial wilderness protection (4% of all land in the United States), far less than 1% of coastal waters in the United States are currently being protected in marine reserves.

▌ Over the past two decades, data from marine reserves around the world have been indicating that reserves can benefit both fish species and fishing economies. On Georges Bank, once commercial trawling was halted in 1994, many marine species began to recover, including benthic invertebrates such as scallops and groundfish such as cod, haddock, and yellowtail flounder. Moreover, fish from protected areas also appear to spill over into adjacent waters, increasing the productivity and overall health of more extensive environments than the initial targeted areas. Important questions remain as to the best design for future marine reserves—their size, number, and placement—to take advantage of ocean currents and fishery protection.

▌ Scientific studies estimate that between 10 and 65% of the ocean's area should be protected as no-take reserves, with most agreement falling between 20 and 50%. This approach would provide sustainable fisheries and healthy fish stocks worldwide.

## *Multiple-Choice Questions*

1. Which of the following would be the *best* indication that a lake was contaminated by human sewage from local septic systems?
   (A) lack of hypoxic zones
   (B) low BOD and a fecal coliform count of 2 colonies/100 ml of sample
   (C) high levels of nitrates and arsenic and a high BOD
   (D) a low BOD and no detectable arsenic
   (E) a high BOD and a fecal coliform count of 5 colonies/100 ml of sample

2. Which of the following is *not* a non-point source pollutant?
   (A) fertilizer runoff from golf courses
   (B) a sewage treatment plant effluent discharged into the local river
   (C) sediment loading of a branching tributary of the Mississippi River
   (D) animal wastes from feedlots
   (E) stormwater runoff from a residential area

3. Ocean acidification is caused by
   (A) dissolving of the shells of marine invertebrates.
   (B) increased oxygen content around coral reefs.
   (C) wastes from the high marine biodiversity found in coral reef ecosystems.
   (D) increased carbonate ion concentration resulting from increased carbon dioxide concentration in oceans.
   (E) increased bicarbonate ion concentration resulting from increased carbon dioxide concentration in oceans.

4. Eutrophication and hypoxia in Chesapeake Bay are in part attributed to
   (A) loss of kelp to take up nutrients.
   (B) loss of otters to maintain the kelp forests.
   (C) overfishing cod that maintain the entire ecosystem.
   (D) overharvesting oysters that previously would have filtered nutrients.
   (E) the recent offshore oil drilling platform accident that contributed excess organic material to this estuary.

5. A eutrophic pond on the grounds of a remote environmental nature center in Colorado would be a prime location for studying
   (A) the effects of heavy metals on aquatic invertebrates.
   (B) the effects of salinization on groundwater.
   (C) the effects of pesticide runoff from the nearby wilderness area.
   (D) the effects of excess nutrients on endemic aquatic algae.
   (E) the effects of aquatic invertebrates on native mammals.

6. In a municipal water treatment plant, the secondary treatment method consists of
   (A) aeration basins, which increase microbial action.
   (B) allowing sewage to sit in settling tanks so that suspended solids will settle out.
   (C) filtering the sewage as it arrives from the influent pipes.
   (D) chlorinating the sewage as it leaves the effluent pipes.
   (E) removing the sludge for the final incineration process.

7.  Two *primary* reasons for the current drought conditions in California and in other parts of the American Southwest are
    (A) excessive lawn-watering and car-washing significantly lowering local water tables.
    (B) global climate change and recent *El Niño* effects.
    (C) forest decline combined with increased agricultural productivity.
    (D) increased ocean temperatures and decreased precipitation.
    (E) record-low precipitation and record-high temperatures.

8.  Scientists are able to predict the possibility and length of *future droughts* and climate change by studying
    (A) rate of marine iceberg melting.
    (B) carbon dioxide levels in ocean waters.
    (C) global forest decline.
    (D) tree ring data from U.S., Canada, and Mexico.
    (E) temperature and precipitation data from the last 100 years in the American Southwest.

9.  Which of the following has *most* contributed to global aquifer contamination?
    (A) inappropriate use of household detergents containing phosphates
    (B) saltwater intrusion from depleted aquifers
    (C) methane and carbon monoxide gases dissolving into aquifers
    (D) septic systems leaching into aquifers
    (E) hazardous waste disposal from pumping waste underground

10. If you wanted to establish an oligotrophic pond at your school's outdoor environmental laboratory, you would want to
    (A) avoid increasing the oxygen saturation.
    (B) decrease the calcium available to the bottom-dwelling organisms.
    (C) avoid runoff from your school's football field maintained by the local lawn-care provider.
    (D) provide carbon-based fertilizers to the entire pond in an effort to increase algal biodiversity.
    (E) provide nitrogen for good nutrient mixing and algal growth stimulation.

11. *Bycatch* refers to
    (A) fishing for two species of fish concurrently.
    (B) the accidental harvesting of animals.
    (C) fishing only at the surface of the ocean.
    (D) fishing in both the surface and deep waters of the oceans.
    (E) the effects experienced after a moratorium is put in place to protect marine species.

*Use the following graph to answer questions 12 and 13.*

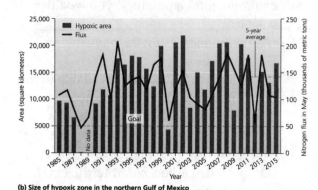

**(b) Size of hypoxic zone in the northern Gulf of Mexico**

**Figure 1b** Area of hypoxic zone in the northern Gulf of Mexico

12. In the year that the nitrogen flux was the highest, what was the *value* of this flux?
    (A) 5,000
    (B) 20,000
    (C) 22,000
    (D) 210,000
    (E) 220,000

13. In what year was this area of the Gulf of Mexico the healthiest and able to support the widest biodiversity of fish and invertebrates?
    (A) 1989
    (B) 2000
    (C) 2002
    (D) 2008
    (E) 2012

14. The aquatic invertebrates in the lake at a local park are dying. A scientist from the Department of Environmental Protection is now investigating the problem. She first measures the dissolved oxygen in order to check the
    (A) influence of local acid precipitation.
    (B) possible negative effects of sedimentation.
    (C) possibility of eutrophication effects.
    (D) possibility of bacterial parasitism of the aquatic invertebrates.
    (E) heavy metals concentration of the lake.

15. Red tides are most often caused by
    (A) increased ocean oxygen content.
    (B) increased ocean carbon dioxide content.
    (C) ocean acidification leading to red algal blooms.
    (D) coastal nutrient pollution from nonpoint sources.
    (E) coastal destruction of wetlands leaving red clays exposed to erosion.

16. During the important thermohaline circulation of global ocean current systems, surface currents are
    (A) saltier and colder.
    (B) cold and dense.
    (C) less salty, less dense, and warmer.
    (D) warm and dense.
    (E) driven by winds from north to south.

17. Humans use fresh water primarily for
    (A) mining and industrial processes.
    (B) electrical production.
    (C) agricultural irrigation.
    (D) residential heating, cooking, and plumbing.
    (E) global nuclear power energy production.

*Free-Response Question*

Commercial catches of Atlantic cod have been studied and well documented in the North Atlantic for the past 150 years, because this fish has represented such an important economic livelihood for Canadians. The graph below indicates exactly when industrial trawling began and when the Canadian government imposed the controversial moratorium. Where appropriate, use the graph to answer the following questions.

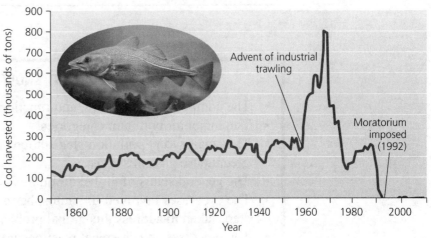

**(a) Cod harvested from Grand Banks**

**Figure 16.23a**

a) Explain, with supporting detail, *two* specific reasons why industrial trawling would produce the change shown in the graph for the 5-year period *after* 1960.

b) Identify the 10-year period during which the greatest decline in the fish harvest took place. For that 10-year period, calculate the average *rate of decline* in the fish harvest, in thousands of tons (metric tons) per year. Clearly show how you determined your answer.

c) Compare and contrast the *environmental effects* of the commercial fishing practices of trawling versus the use of drift nets.

d) Of all the marine pollution effects currently being experienced by the world's oceans, which *two* specific pollutants would marine organisms off the coast of Canada's Grand Banks **most** likely experience? Support your answers with full explanations.

# Atmospheric Science

## *Chapter 17: Atmospheric Science, Air Quality, and Pollution Control*

---

**YOU MUST KNOW**

- The composition, structure, and function of Earth's atmosphere.
- The major air pollutant categories.
- The major air pollution legislation including regulation of greenhouse gas emissions.
- The general status of global air quality.
- The processes involved in stratospheric ozone depletion and the steps taken to address this global problem.
- The processes and chemical reactions involved in acid deposition, including its major environmental consequences.
- The scope of indoor air pollution and its potential solutions.

---

***The composition, structure, and function of Earth's atmosphere***

▌ Earth's atmosphere consists of 78% nitrogen ($N_2$) and 21% oxygen ($O_2$), with the remaining 1% composed of argon (Ar) and minute concentrations of several other permanent gases (remain at fixed concentrations) and variable gases (change in concentration as a result of natural processes or human activities).

▌ Today, human activity is altering the quantities of some of these variable gases, including carbon dioxide ($CO_2$), methane ($CH_4$), nitrous oxide ($N_2O$), ozone ($O_3$), and chlorofluorocarbons (CFCs).

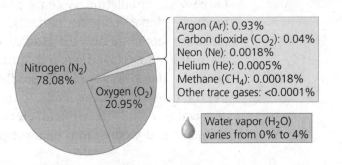

**Figure 17.1** Composition of the Earth's Atmosphere

▌ Earth's atmosphere consists of four layers. The **troposphere** blankets the planet's surface (up to 7 miles high) and provides us with the air we need to live. The **stratosphere** extends from 7 to 31 miles above sea level and contains the **ozone layer**, which greatly reduces the amount of harmful UV radiation reaching Earth's surface. The ozone layer provides a protective layer for life on Earth. The **mesosphere**, extending from 31 to 51 miles above sea level, and the **thermosphere**, extending upward to an altitude of 300 miles, are the two top layers of our atmosphere. Temperature, atmospheric pressure, and relative humidity vary across all four layers.

▌ Normally, temperature decreases as altitude increases in the troposphere. But sometimes a relatively warm layer of air at mid-altitude will cover a layer of cold, dense air below, producing a **thermal inversion**. This warm **inversion layer** traps emissions that then accumulate beneath it and is an air quality factor to consider in many cities. Los Angeles frequently experiences thermal inversions, exacerbating already poor air quality.

▌ Energy from the sun heats the planetary surfaces *unequally* because of latitude differences and the tilt of Earth's axis, causing air and ocean circulation patterns that ultimately produce the seasons and influence weather and climate.

▌ **Weather** specifies atmospheric conditions over short time periods, typically hours or days, and within relatively small geographical areas. Weather is driven in part by the **convective circulation** of air being heated near Earth's surface. This air picks up moisture and rises. Once aloft, the air cools, moisture condenses to form clouds, and precipitation can result. The cooler and drier air descends, beginning the convective cycle anew.

▌ **Climate**, in contrast, describes the pattern of atmospheric conditions found across large geographic regions over long periods of time, typically seasons, years, or millennia.

▮ Large-scale circulation systems consisting of pairs of convective cells produce global climate patterns. Near the equator, intense solar radiation sets in motion the **Hadley convective cells**, resulting in warm air rising and expanding and ultimately releasing the moisture that becomes the heavy rainfall producing the tropical rainforests. The **Ferrel cells** and **polar cells**, two pairs of similar but less intense convective cells, lift air and create precipitation around 60 degrees latitude and dry conditions near the poles.

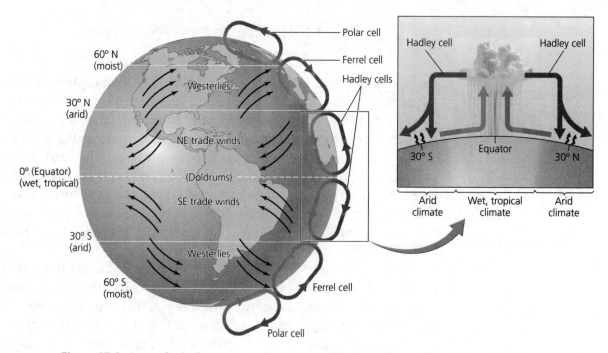

**Figure 17.9** Large-Scale Convective Cells Determine Global Patterns of Humidity and Aridity

### The major air pollutant categories

▮ **Natural sources of air pollution** such as windblown dust, volcanic ash and particulates, and the soot and gases from naturally occurring forest and grassland fires are often made worse by human activity. Land use policies (e.g., "slash-and-burn" agricultural practices) and the build-up of fuel wood from fire suppression in U.S. national forests have increased the amounts of natural air pollutants in our atmosphere. Today, trade winds blow soil across the Atlantic Ocean from Africa to the Americas, harming Caribbean coral reefs with this sediment. Also, strong westerlies lift soil from deserts in Mongolia and China, bringing it across the Pacific Ocean to North America, increasing atmospheric soil particulates.

▮ **Anthropogenic (human-caused) air pollution** can emanate from **point sources** (a specific spot, such as a coal-fired power plant's smokestack) or **non-point sources** (diffuse sources, such as the millions of automobiles spread across a wide area).

▌ **Primary air pollutants** are those coming directly out of a smokestack, exhaust pipe, or natural emission source. They include carbon monoxide, carbon dioxide, sulfur dioxide, nitrous oxides, and most suspended particulate matter. **Secondary pollutants** are often primary pollutants that react with one another or with constituents of the atmosphere (water, $O_2$, sunlight), producing new compounds. Ozone and sulfuric acid are the most prevalent examples.

▌ The U.S. Environmental Protection Agency (EPA), as a result of Clean Air Act legislation, sets standards for six "criteria pollutants," pollutants judged to pose especially great threats to human health. For each of these, the EPA has established **national ambient air quality standards** (NAAQS), which are maximum concentrations allowable in ambient outdoor air. Across the U.S., more than 4,000 monitoring stations take hourly or daily samples in order to assess ambient air quality.

- **Carbon monoxide (CO)** is a colorless, odorless gas produced primarily by the incomplete combustion of fuel. CO poses risk to humans and other animals because it can bind irreversibly to hemoglobin in red blood cells, preventing the essential binding of oxygen.

- **Sulfur dioxide ($SO_2$)** is a colorless gas with a pungent odor. The vast majority of $SO_2$ results from the combustion of coal for electricity generation and industrial processes. Once in the atmosphere, $SO_2$ reacts to form sulfuric acid ($H_2SO_4$), which then returns to Earth as acid deposition.

- **Nitrogen oxides ($NO_x$)** include both NO and nitrogen dioxide ($NO_2$). $NO_2$ is a highly reactive, foul-smelling reddish brown gas that contributes to smog and acid deposition. Most U.S. $NO_x$ emissions result from combustion in vehicle engines.

- **Volatile organic compounds (VOCs)** and mercury, though not officially listed in the Clean Air Act, are commonly monitored by state and local agencies. VOCs are carbon-containing compounds emitted by vehicle engines and used in a wide variety of solvents and industrial processes.

  ▌ One group of VOCs comprises hydrocarbons, such as methane ($CH_4$), propane, butane, and octane.

  ▌ The remainder of VOCs come from natural sources. For example, plants produce isoprene and terpenes, compounds that produce a bluish haze that has given the Blue Ridge Mountains their name.

- **Particulate matter** is composed of solid or liquid particles small enough to be suspended in the atmosphere, including primary pollutants (e.g., soot and dust) as well as secondary pollutants (e.g., sulfates and nitrates). Particulate matter can damage respiratory tissues when inhaled.

▪ **Lead** is a heavy metal that enters the atmosphere as a particulate pollutant. Exhaust from the combustion of leaded gasoline emits airborne lead that can be inhaled or can be deposited on land and water. Lead is a neurotoxin and can enter the food chain, bioaccumulate, and cause central nervous system malfunction.

### *The major air pollution legislation including regulation of greenhouse gas emissions*

▌ **The Clean Air Act of 1970** was a revision of prior Congressional legislation to control air pollution. It set stricter standards for air quality, imposed limits on emissions from new sources, provided new funds for pollution control research, and enabled citizens to sue parties violating the standards.

▌ **The Clean Air Act of 1990** strengthened regulations pertaining to air quality standards, auto emissions, toxic air pollutants (mercury from coal-burning plants and VOCs), acid deposition, and stratospheric ozone depletion. In 1995, an emissions trading program for sulfur dioxide was also introduced.

- This market-based incentive program has helped reduce $SO_2$ emissions nationally.
- Similar cap-and-trade programs for other pollutants, including greenhouse gases, have been implemented.
- Los Angeles adopted its own cap-and-trade program in 1994, decreasing $SO_2$ and $NO_2$ emissions by over 70% by 2010.
- Total emissions of the six "criteria pollutants" have been reduced substantially since passage of the Clean Air Act of 1970.
- These dramatic reductions have occurred despite significant increases in the nation's energy consumption, miles traveled by vehicles, and the gross domestic product. Some of the policy changes and technological developments resulting in cleaner air include the following:

  1) **Catalytic converters** have been required in all new U.S. automobiles beginning in 1975. This required technology converts the hydrocarbons, $NO_x$, and CO emissions into $CO_2$, $H_2O$ vapor, and $N_2$ gas, resulting in cleaner-burning engines.
  2) **Scrubbers** use a combination of air and water that actually separates and removes particulates and $SO_2$ before they are emitted from smokestacks.
  3) Leaded gasoline phase-out caused U.S. lead emissions to plummet by 93% in the 1980s alone.
  4) **Electrostatic precipitators** installed in smokestacks use an electrical charge to make particulates coalesce so that these small particles can be physically removed before entering the atmosphere.
  5) The **sulfur dioxide permit-trading program** and clean coal technologies have reduced $SO_2$ emissions.

▌ **Regulation of greenhouse gases** has been a concern since the 1990s, especially as it relates to global climate change.

▌ In 2007, the U.S. Supreme Court ruled that the EPA has legal authority under the Clean Air Act to regulate carbon dioxide ($CO_2$) and other greenhouse gases as air pollutants; however, this has been challenging to accomplish. In 2011, the EPA introduced moderate carbon emission standards for cars and light trucks. In 2012, it announced it will gradually phase in $CO_2$ limitations for new coal-fired power plants and cement factories.

▌ U.S. $CO_2$ emissions rose by 51% from 1970 to 2005; however, they decreased by 12% from 2005 to 2015 as a result of shifting from coal to cleaner-burning natural gas and from improved fuel efficiency in automobiles.

▌ In 2015, the EPA launched the **Clean Power Plan** for regulating existing power plants by cutting $CO_2$ emissions 32% below 2005 levels by the year 2030.

### *The general status of global air quality*

▌ Some pollutants are not declining, some new air pollutants are emerging, and greenhouse gas emissions continue to rise. Emissions of $CO_2$ and other greenhouse gases that warm the lower atmosphere are arguably today's biggest air pollution problem.

▌ Industrializing nations such as China and India are suffering increasing air pollution problems. For example, China has fueled its rapid industrial development with its abundant reserves of coal, the most-polluting fossil fuel.

▌ In Beijing, China's capital, air pollution became so severe in 2013 that airplane flights were canceled and people were required to wear masks in order to breathe.

▌ The *Asian Brown Cloud*, or *Atmospheric Brown Cloud*, a massive and persistent 2-mile-thick layer of pollution that hangs over southern Asia throughout the dry season, is the result of air pollution primarily from industrialized technologies in China, India, and Iran. China's government is now striving to deal with its severe air pollution problems by closing down some heavily polluting factories; installing pollution-control devices; encouraging development of wind, solar, and nuclear power; and expanding mass transit.

▌ Mexico City has improved its air quality tremendously from its status as the most heavily polluted city in the 1990s. Catalytic converters, Pb removal from gasoline, improved refineries, a mass transit system with a fleet of low-emission buses, and bike-sharing and car-sharing programs have been put in place. As a result, by 2015, most pollutants had been cut by more than 75% and the air was meeting health standards on one in every two days.

- Currently, half the world's urban population breathes air polluted at levels at least 2.5 times beyond the World Health Organization's (WHO's) standards.

- Air quality is also a rural issue worldwide, as a great deal of air pollution emanates from feedlots (dust and the gases of methane, hydrogen sulfide, and ammonia). The worst air quality in the U.S. occurs in rural areas, including California's Central Valley (the nation's agricultural fruit basket).

- **Industrial smog**, or gray-air smog, is the most common air quality problem since the onset of the Industrial Revolution. When coal or oil is burned, some portion is completely combusted, forming $CO_2$, some is partially combusted, producing CO, and some remains unburned and is released as soot (particles of carbon). The sulfur in coal reacts with oxygen to form $SO_2$, which ultimately reacts with water in the atmosphere to also produce sulfuric acid.

- **Photochemical smog**, or brown-air smog, is formed when sunlight drives chemical reactions between the large amounts of NO and VOCs released from the exhaust of automobiles. Over 100 different chemicals are produced, with tropospheric ozone ($O_3$) often being the most abundant. Sunlight is essential for the production of this ozone, and the high levels of $NO_2$ cause photochemical smog to form a brownish haze over cities. This type of smog afflicts many urban areas, especially those with sunny climates and a topography that includes cities surrounded by mountains that trap the pollutants. Los Angeles, despite its progress in reducing Pb and VOCs, still suffers the worst tropospheric ozone pollution of any U.S. metropolitan area according to a 2016 ranking by the American Lung Association.

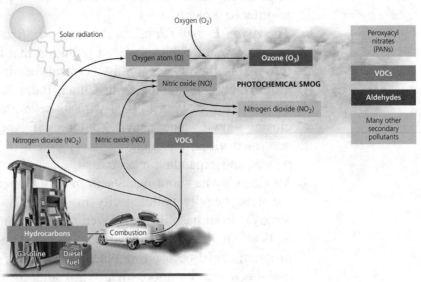

(a) Formation of photochemical smog

**Figure 17.22a** Formation of Photochemical Smog

- Catalytic converter mandates on automobiles, removal of lead from gasoline, use of cleaner gasoline, and removing pollutants from the liquefied petroleum gas that city residents use for cooking and heating are being employed worldwide to reduce photochemical smog.
- Other methods employed include vehicle inspection programs, restricting traffic into city centers, and requiring older vehicles to be taken off the road, with the critical focus on the rising use of automobiles worldwide.

### The processes involved in stratospheric ozone depletion and the steps taken to address this global problem

- **Stratospheric ozone** molecules ($O_3$) are extremely effective at absorbing incoming ultraviolet (UV) radiation from the sun. Even in concentrations of 12 ppm (parts per million), this diffuse concentration of ozone helps to protect all life on Earth's surface from the penetrating and damaging effects of UV radiation.
- **Tropospheric ozone ($O_3$)**, also called **ground-level ozone**, is a colorless gas with an objectionable odor, resulting from the interaction of sunlight, heat, nitrogen oxides, and volatile carbon-containing chemicals. A major component of smog, $O_3$, is a secondary air pollutant that poses human respiratory problems.
- Stratospheric ozone and tropospheric (ground-level) ozone are easily confused, as they are the same gas, $O_3$. However, stratospheric ozone occurs in a *higher* atmospheric layer and performs a *completely different* function.
- Human-made compounds called halocarbons, in which hydrogen atoms are replaced by atoms of chlorine, bromine, or fluorine, have been found to destroy ozone by splitting its molecules apart. **Chlorofluorocarbons (CFCs)** are now known to be one of the specific halocarbons that are **ozone-depleting substances**. These CFCs were mass produced in the 1970s as refrigerants, fire extinguishers, and propellants for aerosol spray cans, and for the manufacturing of polystyrene foams.

  - In the stratosphere, intense UV radiation breaks the bond connecting chlorine to the CFC molecule.
  - Then, in a two-step chemical reaction, each newly freed chlorine atom can split an ozone molecule and then ready itself to split another one. This depletion of ozone molecules results in greater UV penetration of the atmosphere, leading to more skin cancer and, scientists predict, crop damage and reduced productivity of ocean phytoplankton.

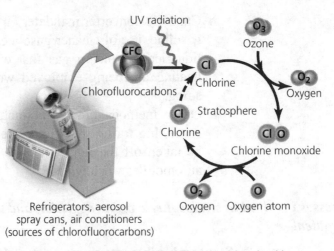

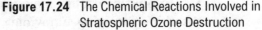

**Figure 17.24** The Chemical Reactions Involved in Stratospheric Ozone Destruction

▮ In 1985, scientists discovered that ozone depletion was most severe over Antarctica, where an "ozone hole" appeared each spring. In response to these scientific concerns, international policy efforts to restrict production of CFCs came together in 1987 with the **Montreal Protocol**. This treaty, with 196 signatory nations, agreed to cut CFC production in half by 1988. Follow-up agreements deepened the cuts, advanced timetables for compliance, and added nearly 100 additional ozone-depleting substances as the industry ultimately shifted to safer alternative chemicals.

▮ Although the Antarctic ozone hole has been stopped from growing worse, it will be many decades before the stratospheric ozone layer is expected to recover completely because of the slow diffusion rate of CFCs into the stratosphere. This fact is one reason scientists often argue for a proactive policy guided by the precautionary principle, rather than reactive policy that risks responding too late.

▮ The Montreal Protocol and its follow-up amendments are widely considered the *biggest success story* so far in addressing *any* global environmental problem and is widely viewed as a model for international cooperation in addressing other global problems from climate change, to persistent organic pollutants (DDT, dioxins, PCBs, etc.), to biodiversity loss.

### *The processes and chemical reactions involved in acid deposition, including its major environmental consequences*

▮ All rain is naturally somewhat acidic because of the reaction between water and atmospheric carbon dioxide, which lowers the pH from a neutral 7.0 to 5.6 because of the presence of the resulting carbonic acid that is produced.

▌ **Acid deposition** refers to the deposition of acidic or acid-forming pollutants that have a pH lower than 5.6 (acid rain, snow, sleet, or hail) and sometimes also consists of dry deposition of acid particulates.

▌ Acid deposition originates primarily with the emission of sulfur dioxide and nitrogen oxides, largely through fossil fuel combustion by automobiles, electric utilities, and industrial facilities. Once airborne, $SO_2$ and $NO_x$ will combine with water to produce sulfuric and nitric acids that may travel days or weeks for hundreds of kilometers before falling in precipitation.

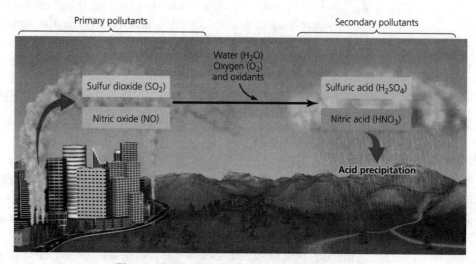

**Figure 17.28**  The Chemistry of Acid Deposition

▌ Acid deposition has wide-ranging, cumulative detrimental effects on ecosystems and on our built environment. These effects include the following.

■ Acids leach nutrients such as calcium, magnesium, and potassium ions from the topsoil, harming plants and soil organisms.

■ Acid precipitation "mobilizes" toxic metal ions such as aluminum, zinc, mercury, and copper by chemically converting them from insoluble forms to soluble forms. These elevated metal ions in soil hinder water and nutrient uptake by plants.

■ When acidic water runs off from land, these metal ions move into water and become deadly to aquatic life by damaging gills, disrupting the salt balance, breathing, and circulation processes.

■ Acid precipitation also damages agricultural crops by damaging leaf surfaces and reducing photosynthetic rates.

■ Stone buildings can be eroded by acid precipitation, reducing ancient and historical architectural treasures.

▌ The severity of all these effects depends not only on the pH of the precipitation or the extent of deposition, but also on the *acid-neutralizing capacity* of the soil, rock, or water that receives the acid input. Precipitation is most acidic in the Northeast and Midwest, near and downwind from areas of heavy industry, with pH readings often ranging from 4.6 to lower than 4.0.

▌ At Dartmouth University's Hubbard Brook Research Forest in New Hampshire, long-term ecosystem studies have taken place since 1963 and individual rainstorms have been measured as low as 2.1—almost 10,000 times more acid than ordinary rainwater.

▌ The 1990 Clean Air Act's emission trading program for $SO_2$ has reduced these emissions, and economic incentives have also encouraged polluters to invest in technologies such as scrubbers. However, in the industrializing world, acid deposition is becoming worse, especially in nations where coal is used widely. China currently has the world's *worst acid rain problem*, as a result of coal combustion in power plants and in factories that lack effective pollution control equipment.

### *The scope of indoor air pollution and its potential solutions*

▌ Indoor air pollution causes far more deaths and health problems worldwide than outdoor air pollution.

▌ In the developing world, indoor burning of fuelwood is the primary risk. Millions of people burn wood, charcoal, animal dung, and crop waste inside their homes for cooking and heating with little or no ventilation.

▌ Even in the developed world, recognizing indoor air pollution as a problem is still quite novel. **Tobacco smoke** and **radon** are the most dangerous pollutants in developed nations. After cigarette smoke, radon gas is the second-leading cause of lung cancer in the developed world, responsible for an estimated 21,000 deaths per year in the United States and 15% of lung cancer deaths worldwide.

▌ Many VOCs are released by everything from plastics to oils to furnishings, building materials, carpets, laser printers, and photocopying machines. With the exception of formaldehyde, the implications for human health of chronic exposure to VOCs are far from clear. However, formaldehyde is one of the most common synthetically produced chemicals, and it has been shown to irritate mucous membranes, induce skin allergies, and cause many other ailments.

▌ Microbes (molds, mildews, and fungi) that induce allergic responses are thought to be a major cause of building-related illness. **Sick-building syndrome** refers to a building-related illness whose cause is a mystery and whose symptoms are general and nonspecific.

▌ Limiting our use of plastics and treated wood as building materials and limiting our exposure to pesticides, cleaning fluids, and other known indoor toxicants are important practices. Most of all, keeping our indoor spaces well ventilated will minimize concentrations of the pollutants among which we live.

# *Chapter 18: Global Climate Change*

---

### YOU MUST KNOW

- Earth's climate system and the many factors influencing global climate.
- How humans influence Earth's atmosphere and climate.
- How scientists study climate change.
- The current and future trends and impacts of global climate change.
- Specific responses and actions we can take to mitigate climate change.
- The economic considerations and approaches involved in the reduction of greenhouse gas emissions.

---

### *Earth's climate system and the many factors influencing global climate*

▌ Earth's climate varies naturally over time and, to some extent, is always changing. However, these changes today are rapidly accelerating, and scientists agree that human activities, notably fossil fuel combustion and deforestation, are largely responsible.

▌ *Climate* describes an area's long-term atmospheric conditions, including temperature, moisture content, wind, precipitation, barometric pressure, and solar radiation.

▌ **Global climate change** describes trends in Earth's climate, involving parameters such as temperature, precipitation, and storm frequency and intensity. Global climate change is *not* synonymous with the term ***global warming***. The latter refers specifically to an increase in Earth's average surface temperature, which is only one aspect of global climate change.

▌ The three factors exerting the most influence on Earth's climate are the sun; the Earth's atmospheric gases that absorb and re-emit the sun's infrared radiation; and the oceans which store and transport heat and moisture.

▌ **Milankovitch cycles** are types of variation in Earth's rotation and orbit that result in slight changes in the relative amount of solar radiation reaching Earth's surface. As the cycles proceed, heat distribution patterns are changed and both ice ages and other climate changes can be triggered.

▌ **Greenhouse gases** in Earth's atmosphere include water vapor, ozone $(O_3)$, carbon dioxide $(CO_2)$, nitrous oxide $(N_2O)$, methane $(CH_4)$, and the synthetic halocarbons. After absorbing radiation emitted from Earth's surface, these gases subsequently re-emit infrared radiation back downward, warming the lower atmosphere (troposphere) in a phenomenon known as the greenhouse effect. Without these gases, our planet would be too cold to support life as we know it. However, it's the **anthropogenic** (human-generated) intensification of the greenhouse effect that concerns scientists. By adding novel greenhouse gases (certain halocarbons) and by increasing the concentrations of several natural greenhouse gases ($CO_2$ and $CH_4$) over the past 250–300 years, humans are intensifying the greenhouse effect beyond what our species has ever experienced.

▌ The **global warming potential** refers to the relative ability of one molecule of a given greenhouse gas to contribute to warming. Values are expressed in relation to $CO_2$, which is assigned a global warming potential of 1. For example, 1 molecule of methane is 84 times as potent as 1 molecule of carbon dioxide.

**TABLE 18.1** Global Warming Potentials of Four Greenhouse Gases

| GREENHOUSE GAS | RELATIVE HEAT-TRAPPING ABILITY (IN $CO_2$ EQUIVALENTS) | |
|---|---|---|
| | OVER 20 YEARS | OVER 100 YEARS |
| Carbon dioxide | 1 | 1 |
| Methane | 84 | 28 |
| Nitrous oxide | 264 | 265 |
| Hydrochlorofluorocarbon HFC-23 | 10,800 | 12,400 |

*Data from IPCC, 2013. Fifth Assessment Report.*

▌ However, since $CO_2$ is the most abundant greenhouse gas in the atmosphere, it contributes more to the natural *and* to the anthropogenic greenhouse effect. According to the latest data, $CO_2$ has caused nearly twice as much warming since the Industrial Revolution, as has methane, nitrous oxide, and halocarbons combined.

### How humans influence Earth's atmosphere and climate

▌ Human activities have boosted Earth's atmospheric concentration of **carbon dioxide** from 280 parts per million (ppm) in the late 1700s to more than 400 ppm today, the highest level by far in over 800,000 years. Increased fossil fuel use (changing ancient terrestrial carbon reservoirs into atmospheric $CO_2$) and increased deforestation (eliminating the capacity of forests to remove current and future atmospheric $CO_2$ during photosynthesis) have both contributed significantly to increased concentrations of $CO_2$ in the atmosphere.

▌ **Methane** concentrations are also rising—150% since 1750, primarily because of livestock metabolic wastes (more feedlots), disposing of more organic matter in landfills, and growing more crops such as rice to feed the world's increasing human population.

▌ **Nitrous oxide** emissions and tropospheric ozone emissions from anthropogenic sources, specifically from increased automobile usage, have also increased 20% since 1750.

▌ **Ozone** concentrations in the troposphere have risen 42% since 1750, but have begun to slow because of the Montreal Protocol.

▌ **Aerosols**, microscopic droplets and particulates (from burning solid materials) can have either a warming or cooling effect, depending on the color of these particles and whether they are located over land or water.

▌ Earth is currently experiencing **radiative forcing** (the amount of change in thermal energy that a given factor causes) of 1.6 watts/m$^2$ of thermal energy above what it was experiencing 250 years ago. This means that today's planet is receiving and retaining 1.6 watts/m$^2$ more thermal energy than it is emitting back into space.

▌ **Ocean absorption** of $CO_2$ is now *less than* the amount of $CO_2$ that humans are adding to the atmosphere. As ocean water warms, it absorbs less $CO_2$ because gases are less soluble in warmer water—a positive feedback effect that *further* accelerates warming of the atmosphere and the potential for climate change.

▌ The pollutants that drive climate change and those that deplete the Earth's stratospheric ozone layer (two separate phenomena!) are each able to cause these global and long-lasting impacts because the pollutants involved persist in the atmosphere for so long.

### *How scientists study climate change*

▌ To have a baseline against which to measure changes happening to our climate today, scientists use **paleoclimate** evidence (data from climate conditions in the geologic past). This evidence is a type of **proxy indicator**, or indirect evidence, that serves as a substitute for direct measurement and can shed light on past climate conditions.

  ▪ **Ice core data**—By examining the trapped air bubbles in ice cores, scientists can determine the atmospheric composition, greenhouse gas concentration, temperature trends, snowfall, and even frequency of forest fires and volcanic eruptions at the time these bubbles of gas were trapped.

  ▪ **Tree ring data**—By examining the width of each tree ring, scientists can determine the amount of precipitation available for plant growth.

  ▪ **Coral reef biochemistry**—Because living corals take in trace elements from ocean water as they grow, their growth bands give clues to ocean conditions at the time the bands were deposited in the coral skeleton.

  ▪ **Sediment core data** beneath bodies of water—Sediments often preserve pollen grains from plants that grew in the past. Because climate influences the types of plants that grow in an area, knowing what plants were present can tell us a great deal about the past climate at that place and time.

▌ **Direct measurements** of temperature, precipitation, and other conditions tell us about current climate and can be compared to the paleoclimate evidence. Direct measurement of $CO_2$ concentrations began in 1958 at the Mauna Loa Observatory in Hawaii and shows an increase from 315 ppm in 1959 to over 400 ppm today.

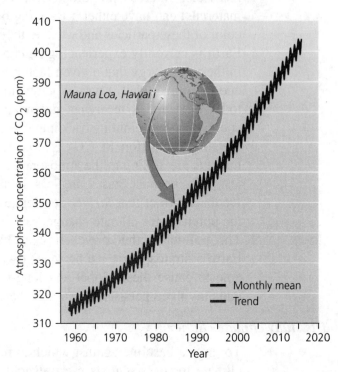

**Figure 18.8** Atmospheric Concentrations of Carbon Dioxide from 1958 to 2016

▌ **Climate models** (programs that combine what is known about atmospheric and oceanic circulation and feedback cycles to simulate climate processes) also serve to predict future changes in climate, in spite of the uncertainties in all their assumptions.

*The current and future trends and impacts of global climate change*

▌ **The Intergovernmental Panel on Climate Change (IPCC)**, an international panel of scientists, has synthesized current climate research data since 1988. This thorough review and widely accepted synthesis of scientific information published from the IPCC has documented observed trends in surface temperature, precipitation patterns, snow and ice cover, sea levels, and storm intensity patterns worldwide. The IPCC was awarded the Nobel Peace Prize in 2007 for its work in informing the world of the trends and impacts of climate change on wildlife, ecosystems, and society, including possible strategies we might pursue in response to climate change.

▌ The 2007 IPCC report concluded that average surface temperatures on Earth increased by an estimated 1.1°C (2.0°F) in the past 100 years, with most of this increase occurring in the last few decades. In the next 20 years, the IPCC predicts that surface temperatures will rise approximately 0.4°C (0.7°F). The IPCC's *Fifth Assessment Report* was released in 2014 and documents further global warming.

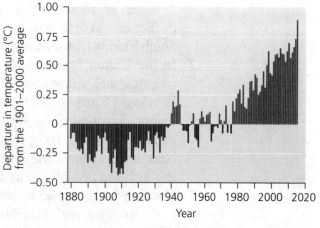
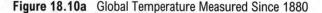

**(a) Global temperature measured since 1880**

**Figure 18.10a**  Global Temperature Measured Since 1880

▌ Average surface temperature on Earth has risen by about 01.1°C (2.0°F) in the past 100 years.

▌ Future **changes in precipitation** are predicted to vary; however, in general, precipitation is expected to increase at high latitudes and decrease at low and middle latitudes, worsening water shortages in many developing countries of the arid subtropics. Extreme weather is now becoming "the new normal."

▌ As Earth warms, mountaintop **glaciers are disappearing**, vast amounts of Arctic sea ice and the Greenland ice sheet are melting, the coastal ice shelves in Antarctica are disintegrating as a result of contact with warmer ocean water, and the Arctic's permafrost (permanently frozen ground) is thawing. The Earth's **albedo** (the capacity to reflect light) is decreasing with less ice, allowing the darker ice-free land to absorb more of the sun's rays, warming these surfaces. This leads to a positive feedback and further warming effects, causing even more snow to melt.

▌ As continental glaciers and ice sheets melt, increased runoff into the oceans causes **sea level rise**. Sea levels also are rising because warmer water has a higher thermal expansion than cooler water. Worldwide, average sea levels rose an estimated 24.1 cm (9.5 in.) in the past 130 years, with an estimated 3.4 mm/year from 1993 to 2016. By 2100, the IPCC predicts mean sea level will be 10–32 inches higher than today's, depending on our future level of greenhouse gas emissions. Sea level rise and climate change have many adverse effects.

1) **Higher sea levels** lead to beach erosion, coastal flooding, intrusion of salt water into aquifers, and the possibility of some low-lying island nations (the Maldives, Tuvalu, Bangladesh, and regions of South Florida and the U.S. Gulf Coast) completely disappearing, displacing hundreds of millions of people. New research since the IPCC's *Fifth Assessment Report* is finding that Greenland's ice is melting at an accelerated rate and that a 1 meter or more rise in sea level by 2100 is a common prediction. At risk will be 180 U.S. cities, including those in South Florida (2.4 million people and 1.3 million homes).

2) In the United States, 53% of the population lives in coastal counties, and vulnerability to **storm surges** as occurred when Hurricane Katrina struck New Orleans (2005) is an ongoing reality. Superstorm Sandy (2012) caused over $600 billion in damages, leaving 160 people dead and thousands homeless in New Jersey. Hurricane Katrina killed more than 1,800 people and inflicted $80 billion in damages.

3) Coral reefs are threatened worldwide, as warming ocean waters leads to coral reef bleaching, ultimately killing corals. Also, enhanced $CO_2$ concentrations in atmosphere are causing more $CO_2$ to be absorbed in oceans, resulting in **ocean acidification** and the potential loss of all shell-building marine organisms, including corals. These effects would reduce marine biodiversity and fisheries significantly because so many marine organisms depend on living coral reefs for food and shelter.

4) As global warming proceeds, it is **modifying biological phenomena** that are regulated by temperature, including spring leaf-out, earlier insect hatching, earlier bird migrations, and earlier animal breeding periods. These shifts can create mismatches between a species seasonal location and the availability of its food source. **Spatial shifts in the ranges** of organisms, with plants and animals moving toward the poles or upward in elevation in response to warmer temperatures, are occurring. However, many species (for example, trees) may not be able to shift their distribution fast enough to avoid extinction.

5) Finally, climate change affects the lives and livelihoods of millions of people because of its **effects on agriculture** (shifting agricultural belts northward may result in lower crop production because of poorer soils at these new latitudes), forestry, and human health (expansion of tropical diseases, increased photochemical smog, and heat stress, causing death).

6) Longer, warmer, drier conditions (along with decades of forest fire suppression) are promoting outbreaks of bark beetles that are destroying millions of trees, **threatening forests** worldwide.

7) The **social cost of carbon** (the cost of damages resulting from each ton of $CO_2$ we emit) has been estimated by the U.S. government to be roughly $40 per ton, and the government determines when and how to impose regulations. The IPCC has estimated that climate change may cost 1–5% of gross domestic product (GDP), with poor nations paying proportionally more.

▌ Climate change and its impacts will vary regionally. For example, in the United States, winter and spring precipitation are projected to *decrease* across the South but *increase* across the North. Sea level rise may affect the East Coast more than the West Coast.

▌ Scientists agree that most or all of today's global warming is due to the well-documented recent increase in *anthropogenic* greenhouse gases in our atmosphere. They also agree that the rise in greenhouse gases results from our combustion of fossil fuels for energy and secondarily from land use changes, including deforestation and agriculture.

### *Specific responses and actions we can take to mitigate climate change*

▌ In the United States, although many of the original naysayers of climate change now admit that the climate is changing, a few still doubt that humans are the cause. However, today in the United States and worldwide, the mainstream debate focuses on how best to respond to the challenges of climate change.

▌ **Mitigation** is the strategy aimed at alleviating the problem; that is, attempting to reduce greenhouse gas emissions so as to lessen the severity of climate change. Examples include energy conservation, improving energy efficiency, switching to clean and renewable energy sources, preventing deforestation, and encouraging practices that protect soil quality. The specific areas in which mitigation strategies are being employed are:

▪ **Electricity generation** is the largest source (40%) of U.S. $CO_2$ emissions. Fossil fuel combustion generates two-thirds of U.S. electricity, and coal accounts for most of the resulting emissions. Encouraging conservation and improved efficiency, as well as making greater use of nuclear power and renewable energy, will reduce these $CO_2$ emissions.

▪ **Transportation** is the second-largest source of U.S. greenhouse emissions. Increasing fuel efficiency for American automobiles and developing technologies that bring alternative approaches (hybrid and fully electric vehicles) to the traditional combustion-engine automobile are reducing greenhouse gas emissions. Development of and a greater reliance on mass transit will be necessary to further reduce transportation's contribution of greenhouse gases.

▪ In **agriculture**, sustainable land management on cropland and rangeland will enable soil to store more carbon. In forest management, the reforestation of cleared areas helps restore forests, which are then able to absorb additional carbon from the atmosphere.

■ Finally, interest in **carbon capture technologies**, which remove carbon dioxide from power plant emissions, is intensifying. **Carbon sequestration** or carbon storage, in which the carbon is stored underground under pressure in rock formations, is being researched; however, this technology is still a long way from being feasible.

■ **Adaptation** is the strategy aimed at minimizing the impacts of climate change by finding ways to cushion oneself from its effects. Examples include restricting coastal development, adjusting farming practices to cope with drought, and modifying water management practices to deal with reduced river flows or salt contamination of groundwater.

■ Both adaptation and mitigation approaches are necessary, because even if we halt all global emissions today, global warming will continue. Mitigation is necessary because if we do nothing to diminish climate change, it will eventually overwhelm any efforts at adaptation we might make.

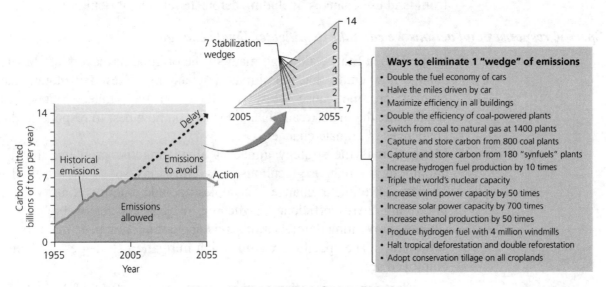

**Figure 18.28** Stabilizing Carbon Emissions

■ Because climate change is a global problem, the world's policymakers have tried to tackle it by means of international treaties. The 1997 **Kyoto Protocol** mandated signatory nations, by the period 2008–2012, to reduce emissions of six greenhouse gases to levels below those of 1990. The United States refused to ratify the Kyoto Protocol and remains the only developed nation not to join this international effort.

■ The United States emits 20% of the world's greenhouse gases, so its refusal to join international efforts to curb greenhouse emissions has generated widespread resentment.

■ As of 2012 (the most recent year with full international data), nations that signed the Kyoto Protocol had *decreased* their emissions by 10.6% from 1990 levels.

■ Nations not a part of the Kyoto Protocol, including China, India, and the U.S., *increased* their emissions by more than 5%.

### *The economic considerations and approaches involved in the reduction of greenhouse gas emissions*

- The U.S. Senate has consistently opposed emissions reductions out of fear that this would dampen the U.S. economy. Industrializing nations such as China and India have so far resisted emissions cuts under the same assumption, given that so much of these economies depend on fossil fuels.
- Germany, Japan, and the United Kingdom, all signatory nations of the Kyoto Protocol and with a standard of living comparable to that of the United States, have significantly lowered their per capita emissions.
  - Germany, the third most technologically advanced economy in the world and a leading producer of iron, steel, coal, automobiles, and electronics, has reduced its greenhouse gas emissions by 25% since 1990.
  - The United Kingdom cut its emissions by 23% during the same time period.
  - Many industrialized nations are positioning themselves to gain economically from major energy transitions in a post-fossil-fuel era. Germany and Japan are currently leading the world in production and deployment of solar energy technology.
  - China recently *surpassed* the United States to become the world's biggest greenhouse gas emitter, although it still emits far less per person than the United States. It is also embarking on initiatives to develop and sell renewable energy technologies on a scale beyond what any other nation has yet attempted.
  - If the United States does not fulfill its potential to develop energy technologies for the future, then the future could belong to nations like China, Germany, and Japan.
- In the absence of action by the U.S. federal government to address climate change, state and local governments across the country are advancing policies to limit emissions.
  - In 2006, California passed the **Global Warming Solutions Act**, which aims to cut California's greenhouse gas emissions 25% by the year 2020.
  - In 2007, 10 northeastern states launched the **Regional Greenhouse Gas Initiative (RGGI)**, a response to set up a cap-and-trade program for carbon emissions from power plants.
- Market mechanisms are being used to address climate change. Supporters of **permit trading programs** (cap-and-trade programs) argue that they provide the fairest, least expensive, and most effective method of reducing emissions. For example, emission sources with too few permits to cover their emissions must find ways to reduce their emissions, buy permits from other sources, or pay for credits through a carbon offset project. Sources with excess permits may keep or sell them, and the price of a permit is meant to fluctuate freely in the market, creating the same kinds of financial incentives as any other commodity in a capitalist economy.

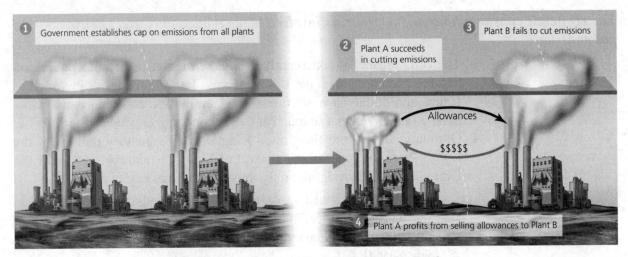

**Figure 18.33**  The Cap-and-Trade Emissions Trading System

- The Chicago Climate Exchange (currently involving over 350 corporations) is the world's *first* (operating since 2003) emissions trading program for greenhouse gas reduction.
- The European Union Emission Trading Scheme, operating since 2005, is the world's *largest* cap-and-trade program; however, it has had some problems with allocating too many permits to their industries, causing the price of carbon permits to fall.

▌ Critics of cap-and-trade systems instead often prefer that governments enact a **carbon tax**. In this approach, governments charge polluters a fee for each unit of greenhouse gas they emit. This gives polluters a financial incentive to reduce emissions; however, many polluters simply pass the cost along to consumers by charging higher prices for their products or services.

▌ A **fee-and-dividend** approach has been put forth by proponents of carbon taxes. In this approach, funds from the tax or "fee" paid to government by polluters are transferred as a tax refund or "dividend" to taxpayers.

▌ **Carbon offsets** are also popular in that they allow participants to make voluntary payments intended to enable another entity to reduce emissions that one is unable to reduce oneself, thus offsetting one's own emissions. For example, a coal-burning power plant might pay a reforestation project to plant trees that will soak up as much carbon dioxide as the coal plant emits.

▌ **Carbon offsets** have become popular among utilities, businesses, universities, and governments trying to achieve **carbon neutrality**, a state in which no net carbon is emitted. These offsets are voluntary payments intended to enable another entity to help reduce the emissions that one is unable to reduce. The payment thus offsets one's own emissions.

▌The most influential economic and environmentally sound factor may be the collective decisions of millions of regular people taking steps to approach a carbon-neutral lifestyle. Determining one's individual **carbon footprint**, the amount of carbon one is responsible for emitting, and then taking the necessary steps to change behavior and consumption choices will be an important factor in addressing the biggest challenge facing current and future generations.

### Multiple-Choice Questions

1. Current average sea level rise rates per year are estimated to be approximately
   (A) 3.4 mm.
   (B) 34 mm.
   (C) 34 cm.
   (D) 250 cm.
   (E) 1 meter.

2. Two major factors involved in the conversion of primary air pollutants into secondary air pollutants are
   (A) sulfur dioxide and sulfuric acid.
   (B) nitrogen oxides and sulfates.
   (C) sulfates and water.
   (D) water and volatile organic compounds.
   (E) sunlight and water.

3. Tropospheric carbon dioxide levels have been directly measured for more than 50 years. Scientists have recently measured these levels at slightly over ___ ppm.
   (A) 300
   (B) 350
   (C) 400
   (D) 415
   (E) 450

4. Of the following greenhouse gases, ___ concentrations have increased the most since 1750, the beginning of the industrial revolution.
   (A) water vapor
   (B) carbon dioxide
   (C) nitrous oxide
   (D) ozone
   (E) methane

5. The pollutant *least* likely to be emitted from an industrial factory's smokestack would be
   (A) carbon monoxide.
   (B) carbon dioxide.
   (C) ozone.
   (D) sulfur dioxide.
   (E) particulates.

6. Under natural, unpolluted conditions, the pH of rainfall is closest to
   (A) 8.0.
   (B) 7.1.
   (C) 5.6.
   (D) 4.5.
   (E) 2.5.

7. The largest *source* of anthropogenic greenhouse gases in the United States is ___, followed by ___.
   (A) agriculture; transportation
   (B) electricity generation; transportation
   (C) electricity generation; agriculture
   (D) electricity generation; industry
   (E) agriculture; electricity generation

8. The Kyoto Protocol
   (A) increased federal funding for controlling greenhouse gas emissions from U.S. power plants.
   (B) required equal concessions from all countries involved in greenhouse gas emission.
   (C) required more concessions from developing countries because of their high dependence on fossil fuels.
   (D) was intended to reduce greenhouse gas emissions to levels lower than those of 1990.
   (E) has not been effective at any level because of the U.S. failure to ratify.

9. Specific examples of the "criteria pollutants" regularly monitored by the EPA are
   (A) $NO_2$, $SO_2$, and particulates.
   (B) $NO_2$, $SO_2$, and $CO_2$.
   (C) $NO_2$, $CO_2$, and VOCs.
   (D) $O_3$, NO, and CO.
   (E) $O_3$, NO, and CFCs.

10. Which of the following is *not* yet regulated by the U.S. EPA?
    (A) carbon monoxide
    (B) ozone
    (C) carbon dioxide
    (D) particulates
    (E) lead

11. Which of the following is a major consequence of acidic deposition?
    (A) It washes toxic metals out of the soil, resulting in healthier soils for biodiversity.
    (B) It results in offshore eutrophication, contributing further to coral reef demise and ocean acidification.
    (C) It leaches out important minerals from soils, resulting in the loss of biodiversity.
    (D) It creates rainwater that can damage skin cells and cause cancer.
    (E) It increases the occurrence of low-lying ground fogs.

12. The greenhouse effect involves warming of Earth's surface and the
    (A) ionosphere.
    (B) thermosphere.
    (C) mesosphere.
    (D) troposphere.
    (E) stratosphere.

13. The Montreal Protocol
    (A) is another example of the many failed attempts to reduce international air pollution.
    (B) is the successful international effort to reduce the effects of global acid deposition.
    (C) developed the much-needed international treaty for eliminating radon emissions from nuclear power plants.
    (D) resulted in significant reduction of CFCs by those countries signing the treaty.
    (E) resulted in significant reduction of carbon dioxide emissions by European countries, Japan, and Australia.

14. The Intergovernmental Panel on Climate Change (IPCC) has synthesized several decades of data and documented observed trends in all of the areas below, *except*
    (A) Earth's surface temperature and its increase.
    (B) the ocean's contribution to photosynthesis and global productivity.
    (C) precipitation patterns and their regional differences.
    (D) snow and ice cover.
    (E) storm intensity patterns.

*Free-Response Question*

Many factors influence Earth's climate. Human activities have also been documented to play a major role. Climate change studies are now the fastest-developing area of environmental science, with new scientific papers and findings refining our understanding of global climate change each month. Today, very few people argue that we need "more proof" or "better science" before we commit to substantial changes involving a new course of action for a sustainable energy policy.

a) Describe/explain three specific *sources of data* that are currently being used to study and document global climate change.

b) The Intergovernmental Panel on Climate Change (IPCC) produced its *Fifth Assessment Report* in 2014, summarizing thousands of scientific studies with specific *observed* trends.. Describe three of these already observed trends that the report has fully documented as occurring. Be sure to include a planetary *physical change*, a *biological effect*, and an effect directly impacting *humans*.

c) Suppose that you would like to make your own lifestyle *carbon-neutral* and that you plan to begin by reducing your greenhouse gas emissions by 25%. What *two* specific actions would you take first to achieve this reduction, and why?

d) Currently the average rate of increase in sea level is 3.4 mm/year. Calculate the expected increase in sea level, in meters, during the next 50 years.

# Energy Resources and Impacts

## *Chapter 19: Fossil Fuels, Their Impacts, and Energy Conservation*

---

**YOU MUST KNOW**

- The nature and origin of each of the major fossil fuel energy sources.
- How to evaluate the extraction process and use of coal, natural gas, and crude oil.
- The nature, origin, and potential of unconventional fossil fuels.
- The ways in which we are extending our reach for fossil fuels.
- The major impacts of fossil fuel use.
- Potential solutions and new approaches involved with fossil fuel technologies.
- The political, social, and economic aspects of fossil fuel use.
- Specific strategies for conserving energy and enhancing efficiency.

---

*The nature and origin of each of the major fossil fuel energy sources*

▌ Since the industrial revolution, nonrenewable **fossil fuels**—including coal, natural gas, and oil—have become our primary sources of energy to heat and light our homes; power our machinery; fuel our vehicles; and produce plastics, pharmaceuticals, synthetic fibers, and more. All fossil fuels are produced by the slow anaerobic (with little or no oxygen) decomposition and compression of organic matter from ancient plant material.

▌ When plants die and are preserved in sediments under anaerobic conditions, they may impart their stored chemical energy (energy originally from the sun that was converted to chemical-bond energy during photosynthesis) to coal, natural gas, or crude oil—each having a high energy content—making them efficient to burn, ship, and store.

▮ Global consumption of the three main fossil fuels has risen steadily for years and is now at its highest level ever. Oil use rose steeply during the 1960s to overtake coal, and today it remains the leading global energy source, representing approximately 35% of the energy consumption market.

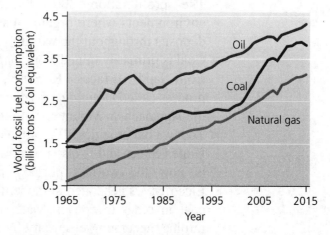

**Figure 19.2**  Global Consumption of Fossil Fuels Over the Past Fifty (50) Years

*Natural gas - Middle East and Russia*

*Coal - United States*

▮ Fossil fuel deposits are unevenly distributed over Earth's surface, with nearly half of the world's proven reserves of crude oil in the Middle East. The Middle East and Russia hold more natural gas than any other country; however, the United States possesses the most coal of any nation.

▮ Developed nations consume more energy than developing nations. For example, the United States has only 4.4% of the world's population; however, it consumes over 18% of the world's energy resources.

▮ To extract, process, and deliver the energy used, it is necessary to invest substantial inputs of energy. **Net energy** expresses the difference between energy returned and energy invested: *Net energy = Energy returned—Energy invested.*

*Net energy = Energy Returned— Energy invested*

▮ The **energy returned on investment (EROI)** ratio is useful when assessing energy sources: *EROI = Energy returned/Energy invested.* Higher ratios mean that we receive more energy from each unit of energy than we invest. Fossil fuels are widely used because their EROI ratios have historically been high. EROI ratios for producing conventional oil and natural gas in the U.S. have declined from roughly 24:1 in the 1970s to around 11:1 today. For the Alberta oil sands (Keystone XL Pipeline project), EROI ratios are even lower, with estimates averaging around 4:1, because oil sands are typically a low-quality fuel that requires a great deal of energy to extract and process.

*EROI = energy Returned / energy invested*

### How to evaluate the extraction process and use of coal, natural gas, and crude oil

■ **Coal** is our most abundant fossil fuel. It results from organic matter (generally woody plant materials) that was compressed under very high pressure to form dense, solid carbon structures. Coal typically results when little decomposition takes place.

- The proliferation 300–400 million years ago of swampy environments where organic material was buried has created coal deposits throughout the world.

- Coal is primarily mined from *underground deposits* and *strip-mined* from land surfaces. Recently, an environmentally destructive process called *mountaintop removal mining* has become common in the Appalachian Mountains of the United States.

- Today, coal provides 40% of the electrical generating capacity of the United States and is also powering China's surging economy. China is now the world's primary producer and consumer of coal. Electricity is generated in modern coal-fired power plants by burning coal in order to convert water to steam. The steam then turns a turbine, generating electricity by passing magnets past copper coils. The steam finally cools and condenses back to water and is returned to the boiler in a continuous loop to repeat the process.

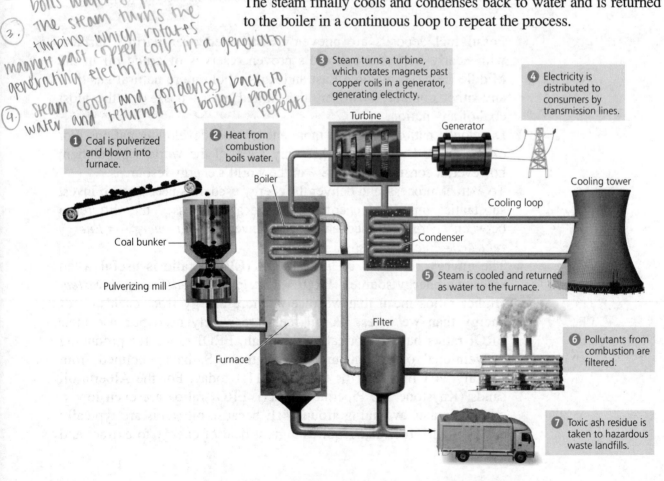

**❸** Steam turns a turbine, which rotates magnets past copper coils in a generator, generating electricty.

**❹** Electricity is distributed to consumers by transmission lines.

**❶** Coal is pulverized and blown into furnace.

**❷** Heat from combustion boils water.

Turbine

Generator

Boiler

Cooling tower

Cooling loop

Coal bunker

Condenser

Pulverizing mill

**❺** Steam is cooled and returned as water to the furnace.

Filter

Furnace

**❻** Pollutants from combustion are filtered.

**❼** Toxic ash residue is taken to hazardous waste landfills.

**Figure 19.10**   Generating Electricity at a Coal-Fired Power Plant

*Handwritten margin notes:*

※ Coal is the most abundant fossil fuel
↓
formed from organic matter

※ Coal Extraction:
mined underground or strip mining

※ Formation of Electricity:
(1.) Coal is burned/heated
(2.) Heat from combustion boils water & produces steam
(3.) The steam turns the turbine which rotates magnets past copper coils in a generator generating electricity.
(4.) Steam cools and condenses back to water and returned to boiler, process repeats

*[Handwritten margin notes:]*
*lignite → most water content*
*Bituminous → more compressed*
*Anthracite → most compressed high potential energy*
*↓ human refined*

- Coal varies in its water and carbon content and the amount of potential energy it contains, ranging from peat (least compressed and most water content), to lignite, sub-bituminous, bituminous, and finally to anthracite coal, the most compressed and containing the highest potential energy.
- Coal deposits also vary in the amount of impurities they contain such as sulfur, mercury, arsenic, and other trace metals.
- Scientists estimate that Earth holds enough coal to supply our society for roughly 110 years, based on the reserves-to-production ratio (R/P ratio), far longer than oil or natural gas.

*[Handwritten margin notes:]*
*★ NATURAL GAS ↓*
*consists of primarily methane and volatile hydrocarbons*
*- "clean-burning" it emits half the amount of $CO_2$ coal produces*

- **Natural gas** consists primarily of methane, $CH_4$, and typically includes varying amounts of other volatile hydrocarbons. It is increasingly favored as a "bridge fuel" because it is versatile and clean-burning, emitting just half as much $CO_2$ per unit of energy produced as coal and two-thirds as much as oil—a "bridge" moving from today's polluting, fossil fuel economy toward a clean, renewable energy economy for the future.

- Natural gas can arise from either of two processes. *Biogenic* gas is created at shallow depths by the anaerobic decomposition of organic matter by bacteria. The decay process taking place in landfills is a source of biogenic natural gas. *Thermogenic* gas results from compression and heat at depths below about 3 km (1.9 mi).

*[Handwritten margin notes:]*
*★ Natural Gas Extraction found above deposits of crude oil or seams of coal*
*↳ Extracted by hydraulic fracturing*
*↓ pipe is drilled into rock formations, special water is pushed along the pipe cracking the rock structure around it and collects bits of natural gas.*

- Most gas extracted commercially is thermogenic and is found above deposits of crude oil or seams of coal, so its extraction often accompanies the extraction of those fossil fuels.
- Today, natural gas extraction also makes use of **hydraulic fracturing** techniques to break into rock formations, as many of the most accessible natural gas reserves have already been exhausted.

*[Handwritten margin notes:]*
*★ PETROLEUM*
*A fluid mixture of hydrocarbons, water, sulfur and natural gas*
*★ Found in underground deposits formed by remains of ocean-dwelling phytoplankton.*
*★ crude oil = liquid petroleum that is a thick mixture of hydrocarbons formed under high temp & pressure*

- **Petroleum** is a fluid mixture of hydrocarbons, water, sulfur, and natural gas that occurs in underground deposits and forms from the remains of ocean-dwelling phytoplankton. **Crude oil** is liquid petroleum that is a thick, liquid mixture of hydrocarbons formed 1–2 miles underground under high temperature and pressure. All crude oil is a mixture of hundreds of different types of hydrocarbon molecules; therefore, the specific properties depend on the chemistry of the organic starting materials, the geologic environment of its location, and the details of the formation process. The U.S. Department of Energy refers to petroleum, crude oil, and oil as interchangeable and roughly equivalent substances.

- Oil is the world's most-used fuel today, accounting for 35% of the world's commercial energy consumption. Geologists infer the location and size of oil deposits by using ground, air, and seismic surveys to map underground rock formations.

*located using seismic activity maps*

*Refining oil*
*① crude oil is heated and boiled*
*② once the hydrocarbons are volatized they proceed through distillation columns*
*③ They are then separated based on different boiling points and weights*

Estimates are then made on the *"technically recoverable"* versus the *"economically recoverable"* amount of oil. The extent to which a company chooses to drill will be determined by the costs of extraction and transportation, balanced against the current price of oil in the world market. The **proven recoverable reserve** is the amount of a fossil fuel that is technologically and economically feasible to remove under current conditions.

- **Oil refineries** separate the many types of hydrocarbon molecules present by using high heat to boil the crude oil, causing its many hydrocarbon constituents to volatilize and proceed through a distillation column. Based on differences in boiling points, the heavier fuels can be separated from the lighter fuels. Combining refining with other chemical processes, crude oil can also be transformed into lubricating oil; asphalts; and the precursors of plastics, synthetic fibers, and pharmaceuticals.

- Some scientists have estimated that we have already extracted half the world's oil reserves. To estimate how long this remaining oil will last, analysts calculate the **reserves-to-production ratio**, or **R/P ratio**. At current levels of production (32 billion barrels globally per year), the estimated remaining 1.7 trillion barrels would last about 53 years. Applying R/P ratios to natural gas reserves has resulted in estimates of a 54-year future.

- **"Peak oil"** is the term used to describe the point of maximum production of petroleum in the world (or for a given nation), after which oil production declines. This is also expected to be roughly the midway point of extraction of the world's oil supplies. Once we pass the peak and production slows, the gap between rising demand and falling supply may pose immense economic and social challenges for our society.

- **Hubbert's peak/Hubbert's curve** was the peak in U.S. production of crude oil. It occurred in 1970, just as Shell Oil geologist M. King Hubbert had predicted in 1956. Predicting an exact date for peak oil *globally* is more challenging because many companies and governments do not reveal their true data on oil reserves. Today, U.S. oil extraction has reached the 1970 amount again, as hydraulic fracking has enabled extraction of formerly inaccessible oil and also the pursuit of various unconventional petroleum sources.

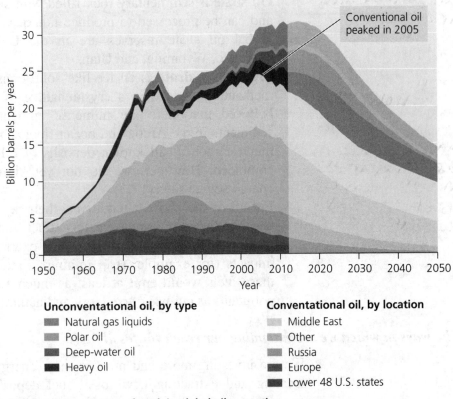

**(b) Modern prediction of peak in global oil extraction**

**Unconventional oil, by type**
- Natural gas liquids
- Polar oil
- Deep-water oil
- Heavy oil

**Conventational oil, by location**
- Middle East
- Other
- Russia
- Europe
- Lower 48 U.S. states

Conventional oil peaked in 2005

**Figure 19.12b**   Modern Prediction of Peak in Global Oil Production

### *The nature, origin, and potential of unconventional fossil fuels*

▪ As oil production declines, we will rely more on natural gas and coal; however, these are also in the process of peaking and declining because of their finite, nonrenewable nature. Only the timing of the "peak" is a question, although most estimates predict a narrow window of time, from now through 2040.

▪ Three types of unconventional fossil fuels exist in large amounts: oil sands, oil shale, and methane hydrate.

▪ **Oil sands** (also called **tar sands**) are deposits of moist sand and clay containing 1–20% bitumen, a thick and heavy form of petroleum that is rich in carbon and poor in hydrogen. Bitumen is too thick to extract by conventional oil drilling, so oil sands are generally removed by strip mining vast areas of land, resulting in mile-wide open pits. For deeper deposits, injecting steam or chemical solvents to liquefy the bitumen is necessary. Canada's controversial **Keystone XL Pipeline** project is focused on extracting these oil sands. In late 2015, President Obama decided against approving the pipeline, noting that it would not lower gas prices, would not enhance America's energy security, and would undercut U.S. leadership in the area of global climate negotiations.

- **Oil shale** is sedimentary rock filled with kerogen (organic matter) and can be processed to produce liquid petroleum. About 40% of global oil shale reserves are in the United States, mostly in Colorado, Wyoming, and Utah.

- **Methane hydrate** is an ice-like solid consisting of molecules of methane embedded in a crystal lattice of water molecules. It is believed that there are immense amounts of this resource in sediments in the Arctic and ocean floor, possibly holding twice as much carbon as all known deposits of oil, coal, and natural gas combined. However, we do not yet know how to extract this energy source safely.

- For all of the above alternatives, their net energy values are low because they are expensive to extract and process—their EROI ratios are low. Also, these fuels exert severe environmental impacts (for example, strip mining). Finally, our combustion of these fuels would emit at least as much $CO_2$, $CH_4$, and other air pollutants as our use of coal, oil, and natural gas.

### The ways in which we are extending our reach for fossil fuels

- We are investing more and more money, energy, and technology into locating and extracting new fossil fuel deposits, with the result of expanding our proven reserves. However, the reality of reducing the EROI fuel ratios, intensifying pollution, and worsening climate change is a major consideration.

- Currently, the major ways of extending fossil fuel extraction and reserves are:

1) **Mountaintop mining for coal:** In this method, entire mountaintops are blasted away to access coal seams. Although it may be economically efficient, it can cause large amounts of rock and soil to slide down slope, resulting in polluting and burying of existing streams and disrupting life for people living nearby.

2) **Hydraulic fracturing for oil and natural gas:** Fracking is a process that injects large volumes of chemically treated water and sand under high pressure into deep deposits of shale in order to fracture them. After the water-chemical mixture is pumped out, the natural gas liberated flows into the well and can be pumped to the surface. Recent studies are showing that the contaminated "produced water" resulting from the fracking process is full of chemical residues that can easily compromise underground water quality and quantity.

3) **Secondary extraction:** Since a typical oil or gas well extraction facility may leave as much as two-thirds of a deposit in the ground after **primary extraction**, companies may return later to use new technologies involving solvents, steam, and water,

*[Handwritten margin notes: Hydraulic fracturing is an example of secondary extraction → Force oil out with high pressures of chemical water.]*

to force the remaining oil or gas out by pressure. This occurs only when market prices are high enough to make this approach profitable. Hydraulic fracturing is an example of secondary extraction.

4) **Offshore drilling in deep waters:** Roughly 35% of the oil and 10% of the natural gas extracted in the U.S. today comes from deposits located offshore, primarily in the Gulf of Mexico and off the southern California coastline. Drilling in shallow water has occurred for several decades; however, as oil and gas reserves have declined and technology has improved, the U.S. long-standing moratorium on offshore drilling was lifted in 2008. In 2010, after British Petroleum's *Deepwater Horizon* oil drilling platform exploded and sank, causing the largest oil spill in world history, the Obama administration canceled all offshore drilling projects, only to backtrack in 2011 by issuing a five-year plan that opened access to 75% of the oil and gas reserves off Alaska and in the Gulf of Mexico.

5) **Moving into ice-free waters of the Arctic:** As global climate change melts the sea ice covering the Arctic Ocean, new shipping lanes are opening, with many nations and companies vying for a position. So far, Royal Dutch Shell has been the only company to pursue drilling in Alaska waters. After many mishaps, problems, and severe Arctic storms, Shell finally withdrew in 2015 after spending $7 billion in efforts to drill there.

6) **Exploiting new "unconventional" fossil fuel sources:** Although the three sources of "unconventional" fossil fuels—oil sands, oil shale, and methane hydrate—could theoretically supply the world for centuries, their net energy values and EROI ratios are very low and these fuels will exert severe environmental impacts on land and water. However, the major concern is that in burning these fuels, the emission of greenhouse gases in excess of that resulting from our current burning of coal, oil, and natural gas would dramatically worsen air pollution and climate change.

### *The major impacts of fossil fuel use*

*[Handwritten margin notes: CO2 warms our planet; Hydraulic fraction also releases GHG]*

▌ **Climate change:** When we burn fossil fuels, we alter fluxes in Earth's carbon cycle by taking carbon that has been retired into a long-term reservoir underground and releasing it into the atmosphere as $CO_2$. Because $CO_2$ is a major greenhouse gas, its increased concentration warms our planet and drives many changes in our global climate. Carbon dioxide pollution is now becoming recognized as the biggest negative consequence of fossil fuel use.

▪ Vast new sources of greenhouse gases from the fossil fuels emerging from U.S. hydraulic fracturing and Canada's oil sands create serious concerns about their potential effects on global climate change.

- Any time natural gas leaks (during extraction, transport, and processing) into the air, its main component, methane, is added to the atmosphere. Methane is a powerful greenhouse gas, so its leaking worsens global warming and climate change.

■ **Human health:** Fossil fuel emissions also affect human health directly, as respiratory irritants, and coal combustion emits mercury that can bioaccumulate in organisms. Some hydrocarbons, such as benzene and toluene, are also carcinogenic, and crude oil often contains trace amounts of known poisons such as lead and arsenic.

■ **Air pollution:** The combustion of oil in our vehicles and coal in our power plants releases sulfur dioxide and nitrogen oxides, which contribute to industrial and photochemical smog and to acid deposition, both major air pollution problems. Acid deposition exerts many adverse effects on freshwater ecosystems and their aquatic life.

■ **Ocean and coastal pollution:** The mining and extracting of fossil fuels leads to many types of water pollution, including the following.

  ■ Oil pollution of marine environments, such as occurred in the 2010 *Deepwater Horizon* offshore drilling platform explosion. Many bird species, shrimp and fish species, and coastal marsh plants were killed in the world's biggest-ever accidental oil spill.

  ■ Oil from non-point sources of pollution, in particular from roadways and residential areas, gets into waterways and eventually ends up in the ocean with toxic effects.

  ■ Groundwater supplies can become contaminated from underground fuel storage tanks containing petroleum products that leak.

■ **Habitat destruction:** Coal mining devastates natural systems by destroying large swaths of habitat, especially with the new mountaintop removal approach. It also can cause chemical runoff into waterways through the process of **acid drainage**—when sulfide minerals in newly exposed rock surfaces react with oxygen and rainwater to produce sulfuric acid that is toxic to organisms.

■ Oil and gas extraction also modifies the environment during development, with construction of road networks, pipelines, housing for workers, and toxic sludge ponds as only part of the habitat destruction involved. Other environmental concerns include:

  1) Extensive environmental studies have been made in many Alaskan locales to examine the potential effects of oil field development on arctic vegetation, air and water quality, and wildlife (including caribou, grizzly bears, and various birds).

  2) The Arctic National Wildlife Refuge (ANWR) on Alaska's North Slope has been the focus of development debate for decades. Advocates of oil drilling have tried to open its lands for drilling and proponents of wilderness have fought for its preservation.

3) The proposed Keystone XL Pipeline route across the sensitive Sandhills region of Nebraska, migration route to the world's Sandhill crane population, has been a major area of environmental concern.

4) Boreal forest destruction in Alberta, Canada, in order to create the vast open pits necessary to mine the oil sands involved in the Keystone XL Pipeline project.

▌**Groundwater contamination:** Hydrofracking presents risks of groundwater pollution that are not yet completely understood, including chemical contamination (by fracking fluids) and methane contamination of drinking water. Hydrofracking also produces immense volumes of wastewater laced with salts, radioactive elements, and other toxic chemicals that become a serious problem to dispose of.

▌**Aquifer overuse and contamination:** The extraction and transport of oil from Alberta Canada's oil sands/Keystone XL Pipeline project also have the potential of contaminating the continent's largest aquifer (the Ogallala aquifer), which serves as the irrigation source for America's breadbasket and the drinking water for millions of people.

### *Potential solutions and new approaches involved with fossil fuel technologies*

▌**Clean coal technologies** refer to a wide array of techniques, equipment, and approaches that aim to remove chemical contaminants during the process of generating electricity from coal. Among these technologies are scrubbers that remove $SO_2$ and/or $NO_x$ from smokestack emissions or filtering devices used to capture tiny ash particles. **Gasification**, in which coal is converted into a cleaner synthetic gas (syngas) by reacting with oxygen at high temperatures, is also an approach being utilized.

▌Current research efforts focus on **carbon capture and storage (CCS)**, or sequestration, as the burning of even the cleanest fossil fuel pumps huge amounts of $CO_2$ into the air, intensifying the greenhouse effect. The CCS approach consists of capturing $CO_2$ emissions, converting the gas to a liquid form, and then storing it in the ocean or underground in a geologically stable rock formation. The world's first coal-fired power plant to approach zero emissions, using carbon capture and storage technology, opened in 2008 in Germany. However, at this point the technology is too unproven to be the central focus of a clean energy strategy for any nation.

▌The costs of alleviating all environmental impacts from fossil fuels are high and the public pays these costs in an inefficient manner. The costs are generally not internalized in the market prices; however, we all pay these external costs through medical expenses, costs of environmental cleanup, and impacts on our quality of life. Equally important to understand is that U.S. fossil fuel prices have been kept inexpensive as a result of government subsidies to extraction companies and fossil fuel industries.

## *The political, social, and economic aspects of fossil fuel use*

▌ Today's societies are so reliant on fossil fuel energy that sudden restrictions on oil supplies can have major economic consequences.

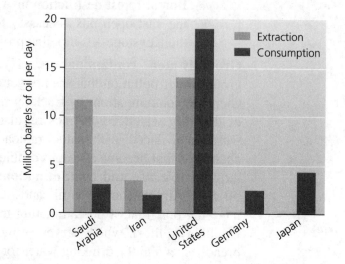

**Figure 19.22**   Comparison of Oil Consumption and Production Patterns

▌ Since its 1970 oil production peak, the United States has relied more and more on foreign energy but today imports just over 20% of its crude oil. This represents a significant reduction since 1970 as conservation measures, increased renewable energy research, secondary extraction, expansion of domestic offshore drilling, and hydraulic fracturing efforts have increased.

▌ In response to the 1973–74 OPEC (*Organization of Petroleum Exporting Countries*) embargo, the U.S. government enacted a series of policies designed to reduce reliance on Middle Eastern oil and moved towards developing additional domestic sources (Alaska's North Slope). Concern over reliance on foreign oil has repeatedly driven the debate to open the Arctic National Wildlife Refuge (ANWR) to drilling, despite critics' charges that such drilling would do little to alleviate the nation's dependence. It is currently estimated that ANWR contains approximately 7.7 billion barrels of "technically recoverable" oil—just over 1 year of U.S. consumption.

▌ Diversification of oil sources towards non-Middle Eastern nations (Canada, Mexico, Venezuela, and Colombia) has been an active approach for the United States. However, even with this approach, there is no guarantee that these nations will continue exporting to the United States over the long term.

▌ The U.S. government is currently urging oil companies to pursue secondary extraction at sites shut down after primary extraction had ceased being cost-effective.

▌ Globally, people living in areas of fossil fuel extraction do not always benefit from extraction of these energy resources, unlike the residents of Alaska who are continuing to receive payouts from the profits of the trans-Alaska pipeline construction in the 1970s and from the ongoing petroleum extraction on Alaska's North Slope.

▌ Today, experts attribute the recent fall in global oil prices to increased supply from hydraulic fracturing in the United States, followed by the decisions of Saudi Arabia and OPEC to boost their own production. For U.S. leaders, the troubled history of the Middle East has enhanced the allure of using hydraulic fracturing to increase domestic oil and gas extraction. U.S. proponents of the Keystone XL pipeline also used this reasoning to argue for importing petroleum from the oil sands of Canada.

▌ In short, fossil fuels are not a sustainable long-term solution to our energy needs since they are all finite in supply and their use has concretely demonstrated adverse health, environmental, political, and socioeconomic consequences.

### *Specific strategies for conserving energy and enhancing efficiency*

▌ Until our society transitions to renewable energy sources, we will need to find ways to minimize the expenditure of energy from our dwindling fossil fuel resources.

▌ **Energy efficiency** describes the ability to obtain a given result or amount of output while using less energy and is generally a result of technological improvements.

▌ **Energy conservation** describes the practice of reducing energy use and results from behavioral choices.

▌ Immediately after the OPEC embargo of 1973–74, the United States enacted a mandated reduction in the national speed limit to 55 mph and increased the mile-per-gallon (mpg) fuel efficiency for newly constructed automobiles. However, these initiatives were abandoned over the next three decades and U.S. policymakers repeatedly failed to raise the *corporate average fuel efficiency (CAFE) standards*, which set benchmarks for auto manufacturers to meet.

▌ Currently, more than two-thirds of the fossil fuel energy we use is simply lost (as waste heat) in power plants and in automobiles. Specific strategies for conserving energy and enhancing efficiency include the following.

　▪ The recent rise in United States fuel efficiency occurred after Congress passed legislation in 2007 mandating that automakers raise average fuel efficiency to 35 mpg by 2020. However, in addition to keeping its taxes on gasoline extremely low, relative to other nations, the United States also lags behind the automobile efficiency of most other developed countries. U.S. gasoline prices do not account for substantial *external costs* that oil extraction and consumption inflict on the environment in the form of land degradation, air and water pollution, and climate change.

■ **Individual behavioral choices** can be made to conserve energy, including driving less, turning off lights when rooms are not being used, dialing down one's home thermostat, and cutting back on the use of energy-intensive machines and appliances. Specific personal actions that promote energy conservation and enhance efficiency include:

1) better home design (passive solar) and increased home insulation.

2) reduction of "vampire" power loss by powering-off electrical appliances and not leaving the appliance in standby mode.

3) use of energy-efficient lighting (compact fluorescent light bulbs, *not* incandescent), which can reduce energy use by up to 80%.

4) use of energy-efficient appliances (refrigerators, dishwashers, etc.) rated by the U.S. EPA Energy Star program.

5) choosing public transportation whenever possible. Alternatively, one can choose electric cars, electric/hybrids, plug-in hybrids, or vehicles that use hydrogen fuel cells. The current U.S. fuel-efficient models obtain fuel economy ratings of up to about 50mpg—twice that of the average American car.

■ As a society, we can conserve energy by making our energy-consuming devices and processes more efficient (using more Energy Star appliances and completely phasing out incandescent lighting in favor of compact fluorescent bulbs).

■ **Cogeneration**, in which excess heat produced (during the generation of electricity) is captured and used to heat nearby workplaces and homes, can be employed to effectively reduce heating requirements.

■ Improved home design and increased insulation are changes that can reduce the energy required to heat and cool residential facilities.

▌ Although essential, effective energy conservation practices and using more efficient technology do not add to our supply of available fuel. Regardless of how much we conserve, we will still need energy, and the only sustainable way of guaranteeing a reliable long-term supply is to ensure sufficiently rapid development of renewable energy sources. The energy path that is chosen will have far-reaching consequences for human health, for Earth's climate, for our environment, and for the stability and future progress of our civilization.

# Chapter 20: Conventional Energy Alternatives

---

> **YOU MUST KNOW**
>
> - The reasons for seeking energy alternatives to fossil fuels.
> - The contributions to world and U.S. energy production by conventional alternatives to fossil fuels.
> - The specifics of nuclear energy, including how it is generated.
> - The specifics of the debate over nuclear power.
> - The major sources, scale, and impacts of bioenergy.
> - The scale, methods, and impacts of hydroelectric power.

## The reasons for seeking energy alternatives to fossil fuels

▌ The global economy is largely powered by fossil fuels, with over 80% of our energy coming from oil, coal, and natural gas.

▌ All of these fossil fuels are finite, nonrenewable energy sources, which contribute to many adverse human health and environmental impacts, the most important of which is the production of greenhouse gases.

▌ In addition to addressing the nonrenewability, adverse air pollution, and climate change issues, developing alternatives to fossil fuels has the added benefit of helping to diversify an economy's mix of energy, thus lessening price volatility and dependence on foreign fuel imports.

## The contributions to world and U.S. energy production by conventional alternatives to fossil fuels

*no other*
*⤷ nuclear, biomass*
*hydroelectric*

*≠ renewable*
*solar, wind,*
*ocean, geothermal*

▌ The alternative energy sources that are currently the most developed and widely used are nuclear power, hydroelectric power, and energy from biomass. Each is thought to exert less environmental impact than fossil fuels, but more impact than the "new renewable" alternatives (solar, wind, geothermal, and ocean power). In addition to having potential for renewability, all these alternatives are best viewed as intermediates along a continuum of renewability depending on how they are utilized.

▌ Although nuclear (5%), hydroelectric (2%), and biomass (10%) contribute substantially to the world's energy consumption, fossil fuels still power two-thirds of global electricity generation.

▌ Sweden has demonstrated that it is possible to replace fossil fuels gradually with alternative sources, having decreased its fossil fuel use from 81% to 30% since 1970. Today, nuclear power, bioenergy, and hydropower together provide Sweden with two-thirds of its total energy and virtually all of its electricity.

▌ By comparison, in the United States these three fossil fuel alternatives provide approximately 16% of our total energy consumption.

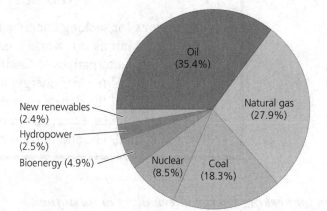

**(a) U.S. energy consumption, by source**

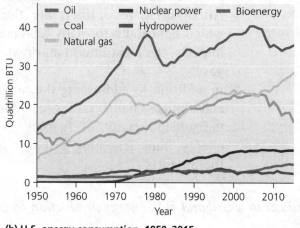

**(b) U.S. energy consumption, 1950–2015**

**Figure 20.3a and b** U.S. Energy Consumption, 1950–2015

## The specifics of nuclear energy, including how it is generated

▌ Nuclear power comes from converting the energy of an atom's nuclear bonds into thermal energy using radioactive elements as the fuel source. **Nuclear fission**, the splitting apart of atomic nuclei, is the reaction that drives the release of nuclear energy in power plants. Uranium-235 is the primary fuel for conventional nuclear reactors, because this isotope is easily split using neutrons, resulting in the production of heat, light, radiation, and additional neutrons, which go on to split additional atoms in a chain reaction.

■ **U-235** is found in uranium ore and represents a very small portion (1%) of the uranium mined. Ninety-nine percent of the uranium from nature occurs as the much less fissionable isotope U-238; therefore, mined uranium ore must be processed to enrich the U-235 concentration to at least 3% in order to sustain a fissionable chain reaction.

■ The enriched uranium is formed into pellets of $UO_2$, which are incorporated into metallic tubes called *fuel rods* that are placed inside the nuclear power plant's *reactor core*. This reactor core is housed within a *reactor vessel*, and the vessel, steam generator, and associated plumbing (carrying cooling water) are often further protected within a meter-thick concrete *containment building*. Not all nations require containment buildings, which points out the key role that government regulation plays in protecting public safety of the global community.

■ In order for fission to begin in a nuclear reactor, the neutrons bombarding U-235 must be slowed down by a substance called a *moderator*, most often water or graphite. Also, excess neutrons produced when uranium nuclei split must be absorbed by *control rods* in order that the fission reaction is maintained at the desired rate.

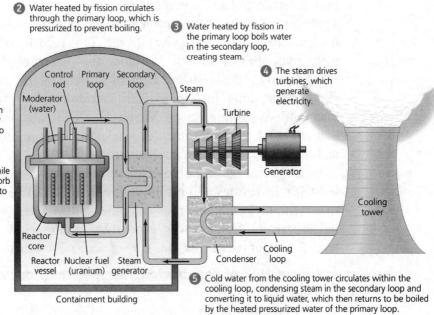

**Figure 20.5** Nuclear Reactors Produce Electricity

■ **Major advantages:** Using fission, nuclear power plants generate electricity without creating the air pollutants (the *most important* of these is the greenhouse gas $CO_2$, as well as $SO_2$ and particulates) generated from fossil fuel emissions. Additional advantages include fewer chronic health risks than are associated with living downwind from power plants burning coal. Also, because uranium generates far more power than coal by weight, less uranium needs to be mined, so there is less damage to landscapes and less solid waste than produced by coal mining.

▪ Today, when considering climate change, the main advantage is that nuclear power avoids $CO_2$ emissions. The International Atomic Energy Agency (IAEA) has estimated that the U.S. avoids emitting 600 million metric tons of $CO_2$/yr, equivalent to the $CO_2$ emissions of almost all passenger cars in the nation and 11% of the total U.S. $CO_2$ emissions.

▪ **Major disadvantages:** The main disadvantage is the problem of safe storage/disposal of the radioactive waste generated and the very long half-life of most nuclear waste isotopes. A second serious consideration is the potential for nuclear plant accidents, as the consequences can be catastrophic because of the radioactive nature of the fuel and waste products from the nuclear fuel cycle. A third consideration is the high cost of building new nuclear power plants and the relatively low EROI value of 14:1 for nuclear power, versus the approximately 45:1 energy return on investment for coal.

### The specifics of the societal debate over nuclear power

▪ Although nuclear power delivers energy more cleanly than fossil fuels, the possibility of catastrophic accidents has caused a great deal of public debate and anxiety over nuclear power.

▪ Three events have been influential in shaping public opinion about nuclear energy: In Pennsylvania, the **Three Mile Island** plant partial meltdown (1979); in the Ukraine, the **Chernobyl** plant explosion (1986); and the **Fukushima Daiichi** nuclear facility accident involving six nuclear units in Japan. This latest accident resulted from the magnitude-9 earthquake on March 11, 2011, and subsequent tsunami that cut off all off-site emergency electricity required for core cooling and on-site waste storage pool cooling.

▪ Even if nuclear power could be made completely safe, we would still be left with the serious and troubling problem, yet to be solved, as to what to do with the spent fuel rods and other radioactive waste. Currently, all nuclear waste from power generation in the U.S. is being held in temporary storage at 120 sites spread across 39 states.

  ▪ Most U.S. plants now have no room left for this type of on-site storage.

  ▪ **Yucca Mountain**, a remote site in the Nevada desert, was chosen as the final repository for U.S. nuclear wastes; however, in 2010 the Obama administration ended support for the project, and an alternative site has yet to be agreed upon.

  ▪ Some citizens voiced concern that the Yucca Mountain site was prone to seismic activity and water infiltration. In addition, the need for highly radioactive wastes to be transported on public highways from the approximately 120 U.S. nuclear facility storage areas has posed additional risks and concerns.

- With all the concerns over waste disposal, safety, and expensive cost overruns, nuclear power's growth has slowed. Electricity from nuclear power today remains heavily subsidized and continues to be more expensive than electricity from coal and other sources.
- However, with increased concern over climate change, carbon emissions, and the Kyoto Protocol agreement, some European nations that had planned to phase out nuclear power have now reversed their decision.
  - Sweden's government has shifted policies from vowing to shut down all nuclear plants by 2010 to now advocating the use of nuclear power as a "bridge" to fight climate change while new renewable energy sources are being developed.
  - The United States halted construction of nuclear plants following the Three Mile Island accident in 1979. At its peak in 1990, the United States had 112 operable plants; today it has 99, with no plans for constructing new nuclear facilities.
  - After the Fukushima accident, the developed countries of Germany, Belgium, and Spain reassessed their nuclear programs and are phasing out nuclear power.

## *The major sources, scale, and impacts of bioenergy*

- **Bioenergy**—also known as **biomass**—is energy obtained from organic matter (biomass) derived from living or recently living organisms. Bioenergy is theoretically renewable and adds no net carbon to the atmosphere because the carbon in biomass fuels may have been captured as recently as months ago (in the case of a corn plant) or perhaps up to a few hundred years ago (in the case of the wood in a tree). Therefore, carbon in biomass is often referred to as *modern* carbon, in contrast to the carbon stored in fossil fuel, which is referred to as *fossil* carbon.
- Bioenergy can come from many different types of plant matter, including wood, charcoal, and agricultural crops, as well as from combustible animal waste products such as livestock manure.
- Fuelwood, charcoal, and manure account for fully one-third of energy used in developing nations; however, in these poorer countries, problems have arisen from overharvesting fuelwood at unsustainably rapid rates, leading to deforestation, soil erosion, desertification, and ultimately diminishing biodiversity.
- New bioenergy strategies, such as burning forestry residues (particularly in Sweden), agricultural residue from cornstalks, animal waste from feedlots, and even organic wastes from landfills (bacterial breakdown of organic matter produces methane, or *biogas*), are being utilized in many industrialized countries today. This latter waste-to-energy process where *biogas* is being produced as a fuel that can be burned to generate electricity is another innovation being employed by Sweden. We are also beginning to grow certain types of plants as crops to generate **biopower**, including fast-growing grasses such as bamboo, fescue, and switchgrass.

▌ **Advantages of biomass:** Biomass power helps to mitigate climate change by reducing $CO_2$ emissions, reduces emissions of $SO_2$ because plant matter (unlike coal) contains no appreciable sulfur content, and enhances energy efficiency by recycling waste products. Capturing landfill gas, to use as a fuel, reduces emissions of methane, a potent greenhouse gas.

▌ **Disadvantages of biomass:** The main disadvantage occurs when crops or plant matter are burned for power, as the soil is deprived of nutrients that would otherwise be returned for future plant growth. The depletion of soil fertility is a major long-term problem for bioenergy and the main reason for its weakness in long-term sustainability.

▌ **Biofuels**, including **ethanol** and **biodiesel**, are used to power automobiles. Ethanol is an energy-rich alcohol produced by fermenting biomass, generally from carbohydrate-rich crops such as corn or sugarcane.

    ▪ About 40% of the U.S. corn crop today is used to make ethanol. At this current level of production, ethanol competes with food production and drives up food costs in many locations.

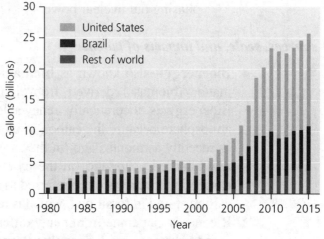

**(b) Global ethanol production, 1980–2015**

**Figure 20.16b** Ethanol Production 1980–2015

    ▪ Growing corn for ethanol also requires substantial inputs of energy, with an EROI ratio of only 1.3:1, meaning that to gain 1.3 units of energy, 1 unit of energy is required.

    ▪ **Biodiesel** is a fuel produced from vegetable oil, used cooking grease, or animal fat. Using waste oil as a biofuel is sustainable; however, most biodiesel today, like most ethanol, comes from crops grown specifically for the purpose (soybeans in Brazil and the United States, oil palms in Southeast Asia, and rapeseed in Europe), all of which take up large areas of land for growing.

    ▪ Because the major crops grown for biodiesel and ethanol exert heavy impacts on the land, research is underway on a variety of other crops from wheat, sorghum, cassava, hemp, and the grass *Miscanthus*.

- Novel biofuels are now being researched and developed. For example, algal cultures grow much faster than terrestrial crops, can be harvested every few days, and produce many times more oil than other biofuel crops. Researchers are also now refining techniques to use enzymes to produce ethanol from cellulose (**cellulosic ethanol**), a substantial advance because cellulose has no food value to people yet is abundant in all plants.
- Theoretically, energy from biomass is **carbon-neutral**, releasing no net carbon into the atmosphere. Burning biomass for energy is *not* carbon-neutral *if* natural forests are destroyed in order to plant bioenergy crops because forests sequester more carbon than do croplands.

## *The scale, methods, and impacts of hydroelectric power*

- **Hydroelectric power**, or **hydropower**, uses the kinetic energy of moving water to turn turbines and generate electricity. Next to biomass, we draw more renewable energy from the motion of water than from any other resource. Hydropower accounts for 2% of the world's energy consumption and 16% of the world's electricity generation.
- Although impounding water from rivers and streams behind dams is the most common approach to generating hydroelectricity, a "run-of-river" system is an alternative approach where water is retained behind a low dam and run through a channel before returning to the river. This latter approach does not store water in a reservoir, as would be the case in a dam. In both approaches, moving water turns the blades of a turbine, which causes a generator to generate electricity.
- Sweden receives 11% of its total energy and nearly 50% of its electricity from hydropower. In hopes of phasing out nuclear power, many Swedish citizens had hoped hydropower could play a still larger role.
- **Advantages:** The main advantages of hydropower over fossil fuels for producing electricity are that it is renewable as long as precipitation falls from the sky, and no carbon compounds are burned in its production of energy. However, fossil fuels *are* used in constructing and maintaining dams and recent evidence indicates that large reservoirs release the greenhouse gas methane as a result of anaerobic decay in deep water.
- In terms of efficiency, hydropower has the highest EROI ratio, averaging 84:1. In other words, for each unit of energy invested in hydropower, 84 units of energy may be returned. Biodiesel, oil sands, and ethanol were at the low end of EROIs, with 2:1, 4:1, and 5:1, respectively. EROIs for fossil fuels were historically very high but have been dropping because we now need to expend more energy to reach less accessible deposits.
- **Disadvantages:** Adverse environmental impacts of hydropower and damming rivers include destruction of habitat for wildlife, disruption of a river's natural flooding cycles, entrapment of sediment behind dams, and the possibility of thermal pollution because water downstream becomes unusually warm if water levels are kept too shallow.

- China's **Three Gorges Dam**, recently completed, is the world's largest hydroelectric plant. Its construction displaced 1.3 million people; however, the dam is now generating as much electricity as dozens of coal-fired or nuclear plants.
- Hydropower may not expand much more because most of the world's large rivers that offer opportunities for hydropower are already dammed. An increased number of people have also become aware of the ecological impacts of dams, and some dams in the United States have actually been dismantled in recent years.
- Although some nations such as Sweden already rely heavily on the use of hydropower, biomass, and nuclear energy, it appears that all countries will need to focus on developing a greater reliance on renewable energy sources in order to move towards a more sustainable energy future.

# Chapter 21: New Renewable Energy Alternatives

> **YOU MUST KNOW**
>
> - The major sources of renewable energy and their potential for growth.
> - The basics of solar energy technology, including both passive and active solar designs.
> - The major advantages and disadvantages of solar energy.
> - The advantages and disadvantages of wind energy.
> - The advantages and disadvantages of geothermal energy.
> - How ocean energy sources are harnessed.
> - The basics of hydrogen fuel cell technology and its potential for growth.

### The major sources of renewable energy and their potential for growth

- The **"new renewable"** energy sources include **solar, wind, geothermal, and ocean energy sources**. These energy sources are "new" in their current rapid stage of technological growth and development; however, they have been utilized by humans for centuries with very simple technology.
- Today, just over 1% of our global energy comes from these "new renewable" energy sources, as compared to the contribution of fossil fuels (81%), nuclear power (8.6%), and the conventional renewable energy sources of hydropower and biomass (10%).

❚ Although they constitute a minuscule proportion of our energy budget, the new renewable sources are growing at much faster rates than all conventional energy sources. The leader in growth is wind power, which has expanded by *nearly 50% each year* since the 1970s.

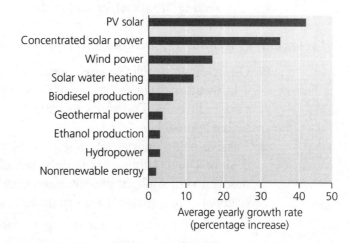

**Figure 21.3** Global Growth of the "New Renewable" Energy Sources between 2010 and 2015

❚ Growing concerns over the diminishing quantity of fossil fuels and their negative impacts on the environment and human health are the main reasons for the growth of wind, solar, and geothermal energy technologies.

❚ Government subsidies have long favored fossil fuels, but new public policies such as the successful use in Germany of **feed-in tariffs** can promote and accelerate our transition to greater use of new renewables.

 ▪ In a feed-in tariff system, utilities are mandated to buy power at guaranteed premium prices from anyone who can generate power from renewable energy sources and feed it into the electric grid. Feed-in tariffs apply to all forms of renewable energy, including solar, wind, and biomass power.

 ▪ For example, German homeowners and businesses have installed more and more **photovoltaic (PV) solar panels** each year, subsequently selling their extra electricity to the utilities at a profit.

 ▪ Germany has now installed nearly half the world's total PV solar technology and obtains more of its energy from solar power than any other nation. Germany also established itself as the world leader in wind power in the 1990s until being overtaken by the United States in 2008 and more recently by China in 2009.

■ Germany's success with feed-in tariffs is serving as a model for other nations in Europe and around the world. As of 2016, more than 90 nations had implemented some sort of feed-in tariffs. In addition, utilities in 46 U.S. states now offer **net-metering**, in which utilities credit customers who produce renewable power and feed it into the electrical grid.

### The basics of solar energy technology, including both passive and active solar designs

■ On average, each square meter of Earth's surface receives about 1 kilowatt of solar energy, which is approximately 17 times the energy of a light bulb. The sun's radiation can be harnessed using passive or active approaches.

■ **Passive solar energy** collection is an approach in which buildings are designed and building materials are chosen to maximize their direct absorption of sunlight in winter and cool the building in summer. Specific features of passive solar energy design can include the following.

■ Installing low, south-facing windows to maximize the capture of sunlight in winter

■ Overhangs to shade these south-facing windows in summer, when cooling, not heating is desired

■ Using construction materials (thermal mass) for floors, roofs, and walls that absorb heat, store it, and release it during the night, when the sun's radiant heat is not available

■ Planting vegetation around a building to buffer or shade the building from temperature swings (evergreens on the north side and deciduous trees on the south)

■ **Active solar energy** collection is an approach in which technological devices are used to focus, move, or store solar energy. These devices are often used to heat water or air in our homes and businesses and include a variety of approaches such as the following.

■ Installing **flat-plate solar collectors** on rooftops that consist of dark-colored, heat-absorbing metal plates mounted in flat, glass-covered boxes. Water, air, or antifreeze run through tubes in these collectors, transferring heat to the building or its water tank for later use. Solar water heating for providing domestic hot water and heating swimming pools is the largest application for this rooftop collector approach in the United States.

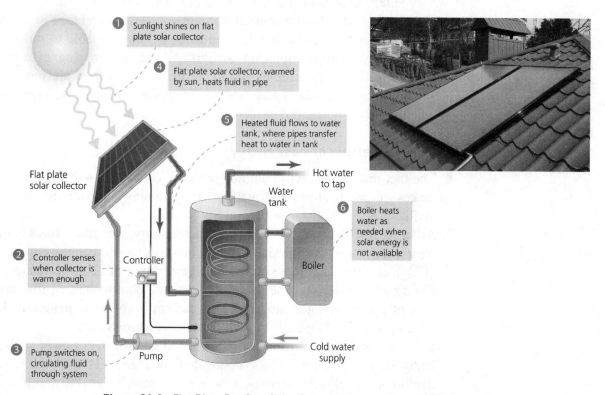

**Figure 21.9**  Flat-Plate Rooftop Solar Collector System Provides Heating

- **Photovoltaic (PV) solar cells** capture energy from the sun as *light*, not heat, and convert it directly into electrical energy. Certain semiconductor materials (silicon) can generate a low-voltage direct electric current when exposed to direct sunlight. This low-voltage is usually converted into higher-voltage alternating current for domestic and business use. The vast majority of PV systems are grid-tied, meaning that any extra electrical energy is sent directly to the utility company, which buys it or gives the homeowner or business credit towards future energy use.

- **Concentrated solar power (CSP)** systems are a large-scale application of solar electricity generation, using mirrors or lenses and a tracking system to focus the sunlight from a large geographical area onto a receiver atop a "power tower." From this point, the CSP system operates like a conventional fossil fuel plant in that the heat from the highly focused beam of concentrated sunlight will heat water (or other liquid) in a steam-driven generator that produces electricity; however, no air pollutants result. The mirror arrays required are large, so CST power plants are best constructed in expansive desert areas where open space and sunlight are plentiful (e.g., Germany's CSP Desertec facility in the Sahara Desert and the Solar Two facility in Southern California).

- **Solar cookers** also use the principle of concentrated solar power to cook food. These simple devices use reflector surfaces in a box-like design to focus sunlight onto food. They are able to raise and maintain temperatures up to 350° F and are proving extremely useful in the developing world. In many places they are replacing the use of already scarce fuelwood supplies, reducing deforestation impacts.

### The major advantages and disadvantages of solar energy

- The fact that the sun will continue burning for another 4–5 billion years makes it practically inexhaustible as an energy source for human civilization.
- **Advantages:** A major advantage of solar energy over fossil fuels is that solar energy does not produce air pollutants, greenhouse gas emissions ($CO_2$), or water pollutants during the *generation* of heat, hot water, or electricity. However, the *manufacture* of photovoltaic cells does currently require water and fossil fuel use and some toxic materials, but once up and running, a PV system produces no emissions.
- Solar systems allow for local, decentralized control over power without being near a power plant or connected to a grid. This is especially helpful in developing nations, where solar cookers enable families to cook food without gathering fuelwood, lessening people's daily workload, and also reducing deforestation.
- The development and deployment of solar systems is producing many new "green collar" jobs. PV technology is the fastest-growing power generation technology today, with China leading the world in yearly production of PV cells, followed by Germany and Japan. The U.S. now accounts for only 2% of the industry's market.
- **Disadvantages:** Location and cost are currently the two major disadvantages to solar energy. Not all regions are sunny enough to provide adequate power; however, the primary disadvantage is the up-front cost of much of the solar technology hardware (other than solar cookers and passive solar home design/orientation).
- Largely because of the lack of the U.S. government's investment or supportive subsidies, currently solar energy accounts for only 0.56% of our energy supply and just 0.94% of the U.S. electricity generation. Even in Germany, which gets more of its energy from solar than any other nation, the percentage is only 6%. However, worldwide solar energy use has grown by 30% annually in the past four decades and it continues to have a large potential for future growth.

### The advantages and disadvantages of wind energy

- **Wind energy**, derived from the movement of air, is really an indirect form of solar energy because it is the sun's unequal heating of air masses on Earth that causes wind movement.

▌ The power of wind can be harnessed by using **wind turbines**, mechanical assemblies that convert wind's kinetic energy into electrical energy. Wind turbines are mounted on towers ranging from approximately 150–350 ft in height. Turbines are often erected in groups called wind parks or wind farms and are located in areas on land or offshore where optimal wind conditions exist.

▌ Denmark is the global leader in obtaining the greatest percentage of its energy from wind power (wind power supplies approximately 40% of Danish electricity needs).

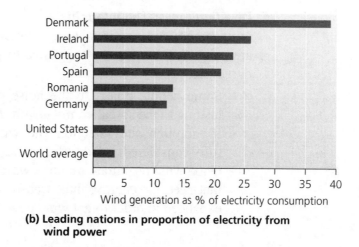

**(b) Leading nations in proportion of electricity from wind power**

**Figure 21.16b**  Leading Nations in Proportion of Energy from Wind Power

▌ Offshore sites hold the greatest promise for future growth because wind speeds are roughly 20% greater over water than over land and there is also less turbulence. Denmark erected the first offshore wind farm in 1991 and over the next decade nine more came into operation across northern Europe where the North Sea and Baltic Sea offer strong winds.

▌ The U.S. **Cape Wind project**, with plans for 130 turbines located off Massachusetts in Nantucket Sound, was given government approval in 2010 after almost a decade of debate; however, construction has yet to begin. As of 2015, 18 more offshore wind developments were in planning stages, mostly off the North Atlantic coast.

▌ **Advantages:** A major advantage of wind is that it is completely nondepletable. The amount of wind available in the future does not depend on how much we use today. Other advantages include the following.

■ Like solar energy, wind power produces no emissions once the equipment (turbine, tower, and infrastructure) is manufactured. The EPA has calculated that running a 1-megawatt wind turbine for 1 year prevents the release of more than 1,500 tons of $CO_2$, 6.5 tons of $SO_2$, 3.2 tons of $NO_x$, and 60 lb of mercury.

- Wind power also appears considerably more efficient than conventional power sources in its energy returned on investment (EROI). The EROI of wind power is 20:1, as compared to a coal-powered electrical power plant's 11:1.
- Unlike conventional fossil fuel or nuclear power plants, wind farms can share the land or offshore waters with other uses, such as grazing of animals, commercial fishing, or recreational activities.
- Wind turbine technology can be used on many scales, from a single tower for local use to fields of hundreds that supply large regions or whole cities.

- **Disadvantages:** The intermittent nature of wind causes unpredictability in its supply and requires storage batteries for residential systems that are off-grid. Batteries are expensive to produce and hard to dispose of or recycle.

  - Some people object to the noise produced by large wind farms, leading to the so-called **not-in-my-backyard (NIMBY)** syndrome.
  - If thorough site selection and wildlife migratory routes are not fully considered, wind turbines may pose a threat to birds and bats.
  - Good wind resources are not always located near population centers that need the energy; thus, transmission networks often need to be expanded in order to get wind power to where people live.

### The advantages and disadvantages of geothermal energy

- **Geothermal energy** is thermal energy that arises from *beneath* the Earth's surface and not from the sun. The radioactive decay of elements under high pressures deep in the interior of our planet generates heat that rises. Where this energy heats groundwater, natural spurts of heated water and steam are sent upwards from below and may erupt at the surface as geysers.
- Northern California, near San Francisco, and in the nation of Iceland are specific locations where geothermal energy is available for development. Geothermal energy can be harnessed directly from these surface geysers, but most often wells must be drilled down at great depths toward the heated groundwater. This heated water can then be used directly for heating homes and offices and for driving industrial processes.
- Geothermal power plants harness the energy of naturally heated underground water and steam to turn turbines and generate electricity. The world's largest geothermal power plants, The Geysers in northern California, provide enough electricity to supply 750,000 homes.
- **Advantages:** A major advantage includes greatly reduced emissions relative to fossil fuel combustion; however, geothermally heated water can release dissolved gases in small quantities, including $CO_2$, methane, ammonia, and hydrogen sulfide.

▌ **Disadvantages:** Geothermal power's major disadvantage is the fact that sources may not always be truly sustainable if a geothermal plant uses heated water more quickly than the groundwater is recharged. Many geothermal plants worldwide are now injecting water back into aquifers to help maintain pressure and sustain the geothermal resource.

▌ **Enhanced geothermal systems (EGS)** are an approach where engineers drill extremely deep into dry rock, fracture the rock, and pump in cold water that then becomes heated. This heated water is drawn up through an outlet well and used to generate power.

▌ **Heat pumps** make use of temperature differences above and below ground. Operating on the principle that temperatures below ground are more stable than temperatures in air, **ground-source heat pumps (GSHPs)** harness the geothermal energy from just below Earth's surface by circulating a fluid through pipes buried underground. In winter, this fluid absorbs heat from underground and then distributes it throughout the home. In summer, when the underground temperature is lower than the air temperatures, the fluid is cooled and pulls heat from the house as it circulates.

### *How ocean energy sources are harnessed*

▌ **Tidal energy** is energy harnessed by erecting a dam across the outlet of a tidal basin. Water flowing with the incoming or outgoing tide through sluices in the dam turns turbines to generate electricity.

▌ The rise and fall of ocean tides twice each day moves large amounts of water past any given point on the world's coastlines. The world's largest tidal-generating facility is South Korea's Sihwa Lake facility, opened in 2011. The La Rance facility in France, which has operated for over 50 years, is the second largest. Many other coastal locations, including San Francisco, New York City, and nations such as China and Canada, are considering further development of their tidal energy potential.

### *The basics of hydrogen fuel cell technology and its potential for growth*

▌ The development of fuel cells and fuel consisting of hydrogen show promise to store energy conveniently, in considerable quantities, and to produce electricity at least as cleanly and efficiently as renewable energy sources.

▌ **A fuel cell** is a device that operates much like a battery where electricity is generated by a reaction between two chemical reactants. Unlike a conventional nickel-cadmium battery, in a fuel cell the reactants are never used up because they are added continuously to the cell, so the cell produces electricity for as long as it is supplied with fuel. Hydrogen is the main reactant in a hydrogen fuel cell and electricity is generated by the process of combining hydrogen and oxygen atoms to form the waste product, water.

▌ Hydrogen gas, usually compressed and stored in an attached fuel tank, is allowed into one side of the fuel cell, whose middle consists of two electrodes that sandwich a hydrogen permeable membrane. On the other side of the cell, oxygen molecules from the open air are split into their component atoms. These oxygen and hydrogen ions then form water molecules that are expelled as waste along with heat.

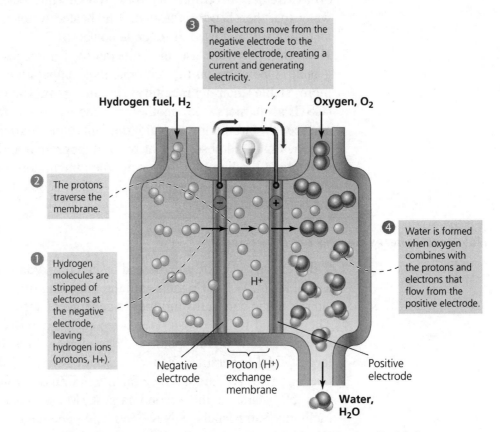

❸ The electrons move from the negative electrode to the positive electrode, creating a current and generating electricity.

**Hydrogen fuel, H₂**

**Oxygen, O₂**

❷ The protons traverse the membrane.

❹ Water is formed when oxygen combines with the protons and electrons that flow from the positive electrode.

❶ Hydrogen molecules are stripped of electrons at the negative electrode, leaving hydrogen ions (protons, H+).

H+

Negative electrode

Proton (H+) exchange membrane

Positive electrode

**Water, H₂O**

**Figure 21.26** Hydrogen Fuel Drives Electricity Generation in a Fuel Cell

▌ The major drawback for fuel cell technology at this point is a lack of infrastructure. To convert a nation like Iceland or Germany to hydrogen as an energy source would require massive and costly development of facilities to transport, store, and provide the hydrogen fuel.

▌ Hydrogen's benefits include the fact that we will never run out of it because it is the most abundant element in the universe. It is clean and nontoxic to use, and, depending on its source, it may produce few greenhouse gases and other pollutants. Pure water and heat may be the only waste products from a hydrogen fuel cell.

▌ Hydrogen fuel cells are energy efficient. Depending on the type of fuel cell, 35–70% of the energy released in the reaction can be used.

❚ Fuel cells are also silent and nonpolluting. For all these reasons, hydrogen fuel cells may soon be used to power cars, much as they are already powering buses operating in some German cities. Hydrogen-powered vehicles could be a sustainable means of transportation because a hydrogen-powered car would use an *electric* motor, which is much more efficient than an internal combustion engine.

❚ The increasing concern over global climate change, air pollution, human health impacts, and security risks resulting from our dependence on finite supplies of fossil fuels have convinced many people that we need to shift to renewable energy sources. Renewable sources with the promise of sustainability include solar, wind, and geothermal energy. Moreover, by using electricity from these new renewables, we will be helping to create a less-polluting global energy market, particularly when global climate change is the major concern and the move toward a more sustainable energy future is the focus.

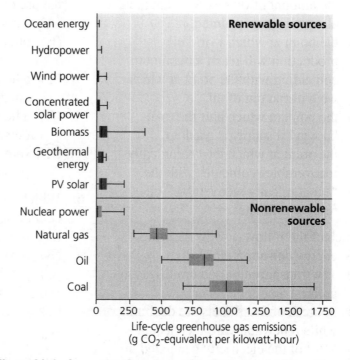

**Figure 21.4** Comparing Greenhouse Gas Emissions for Power Sources

## Multiple-Choice Questions

1. Approximately two-thirds of the fossil fuel energy we use is lost as waste heat, and the transfer of energy from fuels to electricity is about 35% efficient. This is *primarily* a consequence of
   (A) the law of conservation of matter.
   (B) the first law of thermodynamics.
   (C) the second law of thermodynamics.
   (D) the Hubbert curve.
   (E) *CAFE* standards.

2. "Peak oil" is a current topic of global discussion and debate. This topic refers to
   (A) the amount of remaining oil reserves as compared to nuclear fuels.
   (B) the amount of oil reserves that have already been consumed.
   (C) the point at which world oil production will reach a maximum, coinciding with the point at which we will run out of oil.
   (D) the point at which half the total known oil supply is used up.
   (E) the point at which the "technically recoverable" coincides with the "economically recoverable" oil reserve estimates.

3. Which of the following *best* describes U.S. energy consumption?
   (A) The transportation sector utilizes the greatest amount of energy.
   (B) Sources of fossil fuels are declining, while renewables are increasing.
   (C) Most of the electricity generated in the United States comes from nuclear energy.
   (D) Biofuels are currently an important renewable energy resource because of their high EROI values.
   (E) Fossil fuels continue to be the major energy source for all sectors of our economy.

4. Nuclear energy's most *serious* environmental and health concerns focus on the *unresolved* issues surrounding
   (A) the results of a runaway chain reaction.
   (B) the habitat destruction involved in the extensive uranium mining process.
   (C) the short half-life of most of the radioactive isotopes.
   (D) the issue of permanent waste disposal.
   (E) how to pay for the increased costs of this valuable energy source.

5. Which of the following *best* demonstrate(s) the use of passive solar energy?
   (A) a solar cooker
   (B) photovoltaic cells on the roof of an ecologically designed home
   (C) a home with deciduous trees planted outside north-facing windows
   (D) a home with evergreens planted outside south-facing windows
   (E) south-facing windows with stone floors on the inside of the home

6. Which of the following sources of energy *cannot* be connected to the sun's energy?
   (A) biomass
   (B) passive solar energy
   (C) geothermal energy
   (D) wind power
   (E) biofuels

7. For the United States, the *primary* fuel that we use for our energy is
   (A) natural gas.
   (B) coal.
   (C) oil.
   (D) nuclear.
   (E) ANWR petroleum.

8. When we burn fossil fuels,
    (A) we liberate carbon back into the carbon cycle, decreasing the amount available for plant growth.
    (B) the resulting carbonic acid leads to acid rain.
    (C) the greatest environmental impact is the impact on the ozone layer.
    (D) the greatest environmental impact is increased greenhouse gases.
    (E) the greatest environmental impact is the local damage from resource extraction.

9. Which of the following best describes global energy use?
    (A) Electricity generation is powered mainly by nuclear energy.
    (B) Electricity generation is powered mainly by coal.
    (C) Biofuels and biomass energy represent the least $CO_2$ emissions of any energy source.
    (D) Transportation is the largest end use of energy in the world.
    (E) Hydroelectric power represents the lowest EROI for a renewable energy source used globally.

10. Worldwide, what is the most widely used renewable energy resource?
    (A) solar
    (B) hydroelectric        biomass
    (C) biomass
    (D) wind
    (E) nuclear

11. Solar PV-generated electricity as a power source is generated by
    (A) the sun's heat striking solar panels.
    (B) the orientation of the solar panels on a home's rooftop.
    (C) mirrors and lenses that focus the sun's light on a power tower.
    (D) sunlight striking a semiconductor and generating an electrical current.
    (E) the active solar process employed in solar cookers.

12. The primary motivation to develop the new renewable energy sources comes from
    (A) people who believe conservation is essential.
    (B) people who believe that human health is impacted by smog and pollution.
    (C) people who believe fossil fuel dependence has too great a security risk.
    (D) concerns over finite fossil fuels and their environmental impacts.
    (E) concerns over finite food supplies and the need to reverse the practice of growing food for biofuels.

13. In the assessment of energy resources, it is helpful to use a measure called EROI, which is
    E
    (A) energy returned minus energy invested. Net Energy        Energy returned
    (B) energy returned plus energy invested.        Energy invested
    (C) amount of energy invested minus heat released into the environment.
    (D) money invested in extraction and processing minus money in sales.
    (E) energy returned divided by energy invested.

14. Which of the following processes is not a part of the technology of "clean coal"?
    D
    (A) removing sulfur oxides from the emissions
    (B) capturing and sequestering carbon emissions
    (C) converting coal to syngas (natural gas)
    (D) removing carbon content from coal before combustion
    (E) injecting captured carbon dioxide into rock formations deep underground

carbon can't be removed from coal without combustion

15. Which of the following represents the correct ranking of the world's total primary energy sources from the greatest to the least used?
    (A) fossil fuels, biomass, nuclear, hydroelectric, new renewables
    (B) fossil fuels, biomass, hydroelectric, nuclear, new renewables
    (C) fossil fuels, biomass, hydroelectric, new renewables, nuclear
    (D) fossil fuels, nuclear, hydroelectric, biomass, new renewables
    (E) biomass, fossil fuels, nuclear, hydroelectric, new renewables

16. Ultimately, all biofuels are ___ because they are the result of ___.
    (A) geologic; sedimentation
    (B) solar; photosynthesis
    (C) inefficient; decomposition by microorganisms
    (D) synthetic; chemical interactions of hydrocarbons
    (E) organic; chemical metabolic processes of microorganisms

17. Currently, the fastest-growing source of electricity generation using new renewable energy sources is
    (A) passive solar.
    (B) wind.
    (C) hydroelectric power.
    (D) hydrogen fuel cells.
    (E) tidal energy.

18. The Keystone XL Pipeline project refers to the
    (A) removal of coal from Canada's Alberta province.
    (B) extraction of oil from the Gulf of Alaska.
    (C) open pit mining of tar sands in northern Canada.
    (D) hydraulic fracturing of Pennsylvania's Marcellus Shale deposits.
    (E) TransCanada project shipping uranium ore to U.S. nuclear enrichment facilities.

19. Coal continues to be the main fuel to generate electricity in the United States primarily because
    (A) coal causes minimal air pollution when burned.
    (B) coal deposits in the United States lie very close to the surface and are easily extracted.
    (C) compared to natural gas and oil, coal contains very little sulfur, mercury, and arsenic.
    (D) the country has such large reserves of coal.
    (E) the reserves of coal are not found in environmentally sensitive areas.

20. The practice whereby utilities are mandated to buy power from anyone who can generate power from renewable energy sources, passing the excess to the electric grid, is
    (A) net metering.
    (B) referred to as "living off the grid."
    (C) a feed-in tariff system.
    (D) a tax rebate system.
    (E) a renewable subsidy system.

*Free-Response Question*

Although wind power and solar power have been used for centuries as energy sources, people frequently refer to them as "new" renewable energy sources because they are not yet used on a wide scale in modern industrial societies. They continue to be in a rapid phase of development, and they will likely play a much larger role in our future energy use. Photovoltaic (PV) cells, concentrated solar power (CSP) technology, and wind turbines are three of these new renewables receiving much attention. The following table shows the approximate net generation of electricity from wind and solar (PV solar and CSP solar are combined) in the United States from 2011 to 2014 (in million kWh):

| Energy source | Year | | | |
|---|---|---|---|---|
| | **2011** | **2012** | **2013** | **2014** |
| Wind | 120,000 | 140,000 | 168,000 | 182,000 |
| Solar | 7,300 | 12,700 | 21,000 | 33,000 |

(Source: The National Renewable Energy Laboratory, www.nrel.gov; 2014 Renewable Energy Data Book, November 2015, 29)

a) Calculate the average growth rate for both wind and solar electricity generation for the period between 2011 and 2014.

b) Provide two specific explanations for the *difference* in these two growth rates.

c) In the United States, the state of Washington currently has a feed-in tariff system of $0.15/kWh in place for solar PV. This rate would increase to $0.54/kWh if components are manufactured in the state. Explain the feed-in tariff system and its goals.

d) Electricity from wind energy was estimated to cost $0.38/kWh in 1980, and at several U.S. wind farms it is currently $0.03/kWh. Other than this economic advantage, describe two major *environmental advantages* of wind energy including the main reason why wind power continues to represent less than 1% of all U.S. energy sources.

# Part III

# *Answers and Explanations for Topics 1–9*

The answers and explanations for all multiple-choice and free-response questions found in Part II are included in Part III. Each set of questions is organized according to the major environmental science topics discussed in *Environment: The Science behind the Stories*, and can be easily referenced by referring to these nine major text topics on page **iii** of this AP Test Prep Series.

Each type of multiple-choice question that a student would most likely encounter on the AP Exam is included (from traditional questions, to least/most likely questions, classification questions, and the important data interpretation and analysis questions). A concise explanation is provided for each correct response.

One free-response question is also included for each major environmental topic. Because the FRQ always consists of several sub-parts (e.g., a–d), brief explanations for each of these sub-parts are provided. Examples of each of the various types of FRQs (document-based, data-set, and synthesis and evaluation) are included, in order that experience and confidence may be gained with the range of environmental essay writing that a student will encounter on the AP Exam.

# Multiple-Choice Questions

## *Topic 1: Environmental Science and Sustainability*

**1. (D) is correct.** Today it is widely recognized that *sustainability* does not mean simply protecting the environment against ravages of human development, but must also include the triple bottom line: promoting social justice, economic well-being, and environmental quality at the same time.

**2. (C) is correct.** Unaltered ecosystems provide the greatest value; however, once external costs and benefits are factored in, the economic value of sustainably managed ecosystems exceeds that of converted ecosystems, and they have the potential to provide more value for the long term.

**3. (D) is correct.** Currently we are running a global deficit, depleting renewable resources by using them 50% faster than they are being replenished. Comparing the world's 2.7 ha/person ecological footprint with the U.S. average of 7.2 ha/person produces an impact that is 2.7 times greater.

**4. (C) is correct.** Given that the 2005 global footprint was approximately 1.4 (number of planets) and the 1985 global footprint approximately 1.1, the difference is 0.3. Converting this value to a percent (multiply by 100) results in a 30% increase during this 20-year period.

**5. (D) is correct.** Controlled scientific experiments involve manipulating one independent variable at a time. There is no preference for quantitative over qualitative data, and new hypotheses are not always generated; however, the experimental design does involve testing the validity of hypotheses.

**6. (E) is correct.** Ecosystem services are those produced by natural systems, without human intent, but most often with great economic value for humans.

**7. (C) is correct.** Well-designed and controlled scientific experiments must include a testable hypothesis, one independent variable (that which the scientist manipulates), replication, and several quantifiable dependent variables (that which the scientist measures). Although only one independent variable can be manipulated at a time, several parameters can be measured (dependent variables) as a result of this manipulation.

**8. (E) is correct.** A sustainable world culture involves the essentials of reducing overall natural resource consumption and halting population growth—both factors are causative agents for the increasing deterioration of the world's natural resources and systems.

**9. (D) is correct.** An ecological footprint represents the total area of biologically productive land and water necessary to produce the resources needed, and absorb the wastes generated, from a given person or population.

**10. (C) is correct.** Leadership in Energy and Environmental Design (LEED) is the leading set of standards for sustainable building.

**11. (B) is correct.** Environmental science is an interdisciplinary science, which also includes aspects of economics, politics, and the social sciences.

**12. (B) is correct.** By definition, the independent variable is the variable the scientist manipulates/changes during a controlled experiment; the dependent variable is that which is measured and affected by manipulation of the independent variable.

**13. (C) is correct.** Current (2016) world population estimates are closest to 7.4 billion.

**14. (E) is correct.** Forests, water, and soil resources are all potentially renewable. Oil, mineral resources, mineral ores, and natural gas are finite/nonrenewable resources.

**15. (D) is correct.** Systemic approaches involve addressing the root causes of the problem and viewing the "big picture"; whereas a non-systemic approach focuses on the symptoms of the problem, one problem at a time, and is concerned primarily with the short term. Researching and implementing diverse agricultural approaches that rely less heavily on chemical fertilizers is the only "big picture" response of the choices provided. All other responses focus on a single problem/symptom.

**16. (A) is correct.** Carbon-neutrality approaches involve reducing greenhouse gas emissions (specifically $CO_2$) and/or employing carbon offsets, with the effect of adding *no net carbon* to the atmosphere.

**17. (D) is correct.** Water conservation and recycling cost nothing; bicycling is much less expensive than transportation requiring fossil fuel or biodiesel, and consuming local produce also greatly reduces the transportation/fossil fuel costs of shipping food.

## Free-Response Question/Explanation

a) There are several important ecosystem services connected to forests (choices follow). The *primary* ecosystem services provided by forests are 1) nutrient cycling, and 2) reduction of soil erosion. Also, as a part of the global carbon cycle, forests take up $CO_2$ from the atmosphere during photosynthesis, storing it as organic compounds (biomass). This removal of carbon helps to 3) stabilize atmospheric temperatures, slowing projected climate change. By holding soil in place through their extensive root systems, forests keep soils intact so that plant biodiversity is maintained. This intact soil/plant community also aids in 4) water purification, cleansing the water as it moves through the soil profile.

b) Given that the original ecological services valuation in 1997 of forests was approximately \$4.7 trillion, dividing \$4.7 trillion by 4 results in \$1.17 trillion per year as the approximate economic valuation of forests at that time. To check this answer, $1.17 \times 400\% = \$4.7$ trillion per year. In other words, the value of forests in providing *ecosystem services* (\$4.7 trillion) is much greater than the value of forests for the direct economic gain (\$1.17 trillion) of harvesting for wood products.

c) Sustainability is the approach that focuses on conserving the Earth's functioning ecosystems intact for the long-term, taking no more potentially renewable resources at any one time than can be replenished by these fully functioning ecosystems. It also includes not overloading the capacity of the environment to cleanse and renew itself of wastes by natural processes. Therefore, by removing forests for economic gain, the beneficial ecosystem services provided by these forests become altered. For example, the forest's ability to recycle essential nutrients, purify air and water, stabilize soil, and store atmospheric carbon become greatly diminished, resulting in a planet's reduced ability for sustainability.

# Topic 2: Principles of Ecology

**1. (E) is correct.** Almost all organisms use the solar energy stored in food to fuel their life processes. Autotrophs convert solar energy to glucose, an organic compound useful to both plants and animals. At each successive trophic level, less energy is available as energy is always degraded to less useful forms (heat). Energy cannot be recycled.

**2. (A) is correct.** Since the pH scale is logarithmic, each unit change is by a factor of 10. An increase of pH means the solution is more basic and has 100 times fewer $H^+$ ions, or conversely, 100 times more $OH^-$ ions.

**3. (E) is correct.** Half-life is defined as the length of time it takes for one-half of the radioisotope to emit radiation and decay/change. It will then take 30 years to go from 80 lb to 40 lb (one-half the original amount), and will take another 30 years to result in 20 lb, etc. In other words, 7 "half-lifes" are required to result in less than 1 lb. $30 \times 7 = 210$ years.

**4. (E) is correct.** Denitrification is a process carried out by soil bacteria when organisms die, converting $NO_3$ to $N_2$, thereby releasing nitrogen back into the atmosphere and completing the cycle.

**5. (D) is correct.** Synthetic ammonification (Haber-Bosch process) enabled people to overcome the limits on productivity, long imposed by nitrogen scarcity in nature. This synthetic process ($N_2 \rightarrow NH_4$) produced new fertilizers that have been extensively used by modern agriculture and homeowners, dramatically altering the nitrogen cycle. When our farming practices speed runoff and allow soil erosion, nitrogen (often in the form of nitrates) flows from farms into aquatic ecosystems, leading to nutrient pollution, eutrophication, and hypoxia.

**6. (A) is correct.** Nitrogen fixation is carried out by symbiotic bacteria living in the nodules of legumes (soybeans, peanuts). This process changes atmospheric $N_2$ to $NH_3$, thus allowing N to enter the ecosystem in a biologically active form.

**7. (E) is correct.** Specialists succeed over evolutionary time by being extremely good at the things they do, but they are extremely vulnerable when conditions change and threaten the resource on which they have specialized.

**8. (B) is correct.** Climatographs are graphs depicting the annual temperature and precipitation of a biome. The only answer that discusses precipitation is (B); both deserts and tundra have low precipitation, are quite dry, and support less plant life than other terrestrial biomes. Although some high-altitude deserts have seasonally low temperatures, the tundra's temperature is low year-round.

**9. (E) is correct.** Trophic cascades are defined by top carnivores (tertiary consumers), occupying the fourth trophic level. When present, these top predators keep the carnivores in the third trophic level (below them) in check, and thus members of the second trophic level (primary consumers) will increase in numbers, in an effect known as a trophic cascade.

**10. (B) is correct.** In general, eutrophication is a result of over-enrichment by nutrients, primarily nitrates. In the case of the Gulf of Mexico, specifically, since it is connected to the Mississippi River watershed, the primary cause of eutrophication is the nitrate enrichment found in fertilizers used in modern agriculture. Answers (A) and (E) are the *results* of nutrient enrichment, not the cause.

**11. (C) is correct.** Positive feedback loops act to further enhance the particular effect, driving the system toward one extreme or another. If the climate warms enough to melt Arctic ice and snow, darker surfaces of land and water will be exposed, which will then absorb more sunlight, causing even further melting. As a population of prey increases, this expands the food supply for predators, but as the predators increase in number, they catch more prey and the prey decreases, an example of a negative feedback loop.

**12. (C) is correct.** A location where two ecosystems meet (a forest and a coastal marsh) are transition zones, also known as ecotones. An ecotone contains elements of both ecosystems and has higher biodiversity than either of the separate ecosystems.

**13. (B) is correct.** Only the combinations (B), (D), and (E), which culminate in extinction, can be initially considered. However, (D) includes emigration, which cannot influence *global* immigration, and (E) includes extirpation, which affects only *regional* extinction, not global species diversity. (B) involves speciation and extinction, both global processes that can change species diversity.

**14. (E) is correct.** Top predators are most vulnerable because their numbers are fewer and they are usually larger in size, are long-lived, and raise few young in their lifetime—a classic K-selected species. The Siberian tiger, gray wolf, and sperm whale are all examples of top predators vulnerable to human hunting and also to possible biomagnification of anthropogenic pollutants (pesticides, endocrine disruptors).

**15. (E) is correct.** Bioaccumulation is the process in which persistent toxicants (pesticides) can accumulate in an animal's body in a greater concentration than exists in the surrounding environment. Because pesticides are fat-soluble and would be stored in the bat's fatty tissues, rather than being broken down and excreted, the pesticide concentration would increase with time in the bat's body. Bats are not top predators (species that consume other carnivores) or umbrella species (require a very large area of habitat, which when set aside would offer refuge to many other species). Normal consumption of insects by bats will neither reduce insect biodiversity nor cause regional extinction (extirpation) of insects.

**16. (D) is correct.** Normal ecosystem functioning requires that matter always recycle (it can never be destroyed, according to the law of conservation of matter) and energy be able to flow (according to second law of thermodynamics, in all energy transformations between trophic levels, energy is degraded and can never be recycled). All ecosystems require a constant input of energy, usually from the sun.

**17. (B) is correct.** According to the 10% law, assuming an energy content of 500,550 kcal in the autotrophs located in the first trophic level, then organisms in the third trophic level would contain 1/100th of this amount, or 5,005 kcal of energy. At most, only 10% of the energy at any trophic level can be passed up to the next higher level.

**18. (B) is correct.** Cellular respiration is the energy-releasing reaction of living organisms in which $O_2$ is required to break down the glucose molecule, also releasing $CO_2$ and $H_2O$ in the process. $CO_2$ is then released back to the atmosphere, where it can become available to plants as one of the necessary requirements for photosynthesis.

**19. (D) is correct.** Density-independent factors are limiting factors whose influence is not affected by population density. Catastrophic events (floods, fires, etc.) and temperature extremes can eliminate large numbers of individuals without regard to their density. Density-dependent factors increasingly slow population growth as that population's density increases.

**20. (C) is correct.** Introduced species that are successful and become invasive are usually generalists, having a broad niche. They can outcompete endemic species (those native species found in a given area) because the introduced species often do not have predators in their new environment. Successful introduced species are also often r-strategists, having a large number of offspring in a short period of time; they are not K-strategists.

**21. (E) is correct**. An ecological niche reflects a species-specific use of resources and the role of this species in a community. Niche includes habitat and *all* relationships/roles of the species (symbiotic and feeding relationships, to name just a couple).

**22. (B) is correct.** Producers are the source of all usable energy in an ecosystem, as they are able to incorporate the sun's energy into organic molecules through photosynthesis, thus making that energy available to other individuals occupying higher trophic levels. As this energy is passed up food chains, the energy content diminishes by approximately 10% at each trophic level.

**23. (B) is correct.** The Florida Everglades has recently undergone widespread water depletion due to overdrawing of this natural wetlands area for irrigation, development, and flood control. Restoring historical water levels is the goal of this project.

**24. (E) is correct.** When members of the same species are competing for limited resources, the competition is intraspecific. Interspecific competition is between members of different species.

**25. (A) is correct.** Productivity is defined as the total rate of photosynthesis, or the rate at which autotrophs convert energy to biomass. This question is specifically comparing $gC/m^2$ of the various ecosystems, *not* the total area devoted to each of these ecosystems. In terrestrial ecosystems, productivity tends to increase with temperature and precipitation. In aquatic ecosystems productivity tends to rise with light and availability of nutrients. Water is often limiting in grasslands and deciduous forests, nutrients are limiting in the open ocean, and temperature is limiting in boreal forests. In swamps and marshes, there are large amounts of nutrients, water, and light.

**26. (D) is correct.** With the advent of modern fertilizers high in nitrogen (ammonia), humans have doubled the rate of industrial nitrogen-fixation processes. These N fertilizers are often eroded into waterways, producing eutrophic conditions in aquatic systems. In the past, the large reservoir of nitrogen (78% of the atmospheric gases) only slowly moved into terrestrial systems through microbial nitrogen fixation or lightning strikes.

**27. (E) is correct.** The rate at which biomass becomes available to *consumers* is defined as net primary productivity (NPP). Gross primary productivity is the conversion of solar energy, $CO_2$, and $H_2O$ into organic molecules ($C_6H_{12}O_6$) by *autotrophs*. The energy that remains after cellular respiration is net primary productivity: $GPP - R = NPP$.

**28. (D) is correct.** Earthquakes are becoming more frequent in the south central U.S., particularly in Oklahoma, where the pursuit of seismic-inducing activities, such as oil extraction and wastewater well injection, have been documented as increasing dramatically.

**29. (D) is correct.** Percent change is calculated using the following formula:

Final value – Initial value ÷ Initial value × 100 = % Change.

Given that the number of earthquakes in 2015 was 2,750 and the number in 1995 was 500, use these values in the % change equation:

$(2750 - 500) \div 500 \times 100 = 450\%$.

**30. (D) is correct.** The definition of a species is based upon reproductive compatibility of individuals, not only whether they will mate/breed, but also whether individuals are genetically compatible, producing fertile, viable (living) young, which can then carry on the next generation. Similarity of physical appearance is not a criterion, nor is having similar or even the same food requirements.

**31. (A) is correct.** Plate tectonics is the process by which Earth's surface is shaped by the extremely slow movement of tectonic plates, or sections of crust (approximately 15 major plates). Their interaction gives rise to processes that build mountains, cause earthquakes, and otherwise influence the shape of the landscape and thereby the location of living organisms on the planet.

**32. (E) is correct.** Scientific research is showing that trees grown with $CO_2$ + ozone grow less than with $CO_2$ alone. Although minor variations in tree response can exist, overall (E) is the correct response. Initially it was hypothesized that increasing $CO_2$ levels would always promote tree growth (B); however, current findings indicate a more complex relationship between $CO_2$ and ozone, offsetting increased growth.

**33. (B) is correct.** Under ideal conditions, and after initial introduction, food and space are in great supply, and the limiting factors of environmental resistance have not yet been experienced.

**34. (D) is correct.** To obtain an overall population growth rate:

$(200 + 100) - (100 + 100) \div 1000 = 0.1 = 10\%$.

**35. (B) is correct.** Zebra mussels are filter-feeders, consuming phytoplankton and small zooplankton in the water column. The recent decline of zooplankton since introduction of the zebra mussel has been a direct result of predation by zebra mussels.

**36. (D) is correct.** Net Primary Productivity (NPP) is calculated using the following formula: NPP = Gross Primary Productivity (GPP) minus Respiration. Since it is given that in this New England forest respiration represents 30% of GPP ($70$ g/m$^2$/day $\times$ 0.30), respiration would equal 21 g/m$^2$/day. It then follows that NPP = 71 g/m$^2$/day $-$ 21 g/m$^2$/day = 49 g/m$^2$/day.

**37. (C) is correct.** Keystone species are those that have strong or wide-ranging impacts far out of proportion to their actual abundance. They are not dominant in numbers, are never producers, and don't cause removal of a whole trophic level, but they are most likely top predators whose presence can often cause a trophic cascade.

**38. (B) is correct.** Although all answers are true at some level, on a scale of importance, the overdrawing of groundwater for modern agriculture, industrial, and drinking water purposes is considered the most critical. Usable, potable fresh water represents less than 1% of all water on Earth, and these resources are dwindling at an alarming rate.

**39. (C) is correct.** The largest carbon reservoir was deposited over millions of years by dead plants and animals, buried by sediments, and turned into sedimentary rocks and ultimately fossil fuels. The atmosphere and hydrosphere both contain small amounts of carbon (as $CO_2$). The abundance of plants and the fact that they take in so much $CO_2$ for photosynthesis makes them a major reservoir of carbon, but not as large as sedimentary rock.

**40. (D) is correct.** "Dead zones" are hypoxic areas resulting from eutrophication processes that were initiated by nutrient enrichment, usually nitrates in marine ecosystems. Negative feedback loops (A) enhance ecosystem stability and are not characteristic of processes operating in "dead zones." There is no evidence that hypoxic zones are caused by introduced species (B).

**41. (C) is correct.** When oceans become acidic, the pH is *lower*, not higher (B). Ocean acidification *decreases* the availability of carbonate ions necessary for shell-building. The lowering of pH due to increased carbonic acid formation (atmospheric $CO_2$ + $H_2O$ → $H_2CO_3$) has the ability to dissolve invertebrate shells and dramatically decrease productivity of coral reefs; however, it is not the dissolving of shells that *causes* ocean acidification (E).

**42. (D) is correct.** Biodiversity results from the sum of speciation minus extinction. High rates of mutation alone do not necessarily produce greater speciation, as it could be followed by high rates of extinction. A wide variety of habitats does not necessarily mean that all habitats will be filled.

**43. (E) is correct.** Of the choices provided, habitat destruction is the most critical factor in determining species extinction rates. Overharvesting, introduction of invasive species, and global climate change all have the potential to cause species extinction; however, habitat destruction (due to growing human populations and their need for food and energy) is today's number one threat to species.

**44. (D) is correct.** A community is made up of numerous populations of different species in a given area. The ecosystem level would include the community of different populations plus the specific environmental factors (temperature, soil, moisture, light, etc.) in that specific area.

*Free-Response Question/Explanation*

a) *Three characteristics* of a species that would make the species particularly susceptible to extinction are (choices follow): 1) **low reproductive rate** (1–2 offspring, possibly not every year, i.e., K-strategist), 2) **slow maturity** (taking many years to reach sexual maturity; thus any mortality is not easily recouped, 3) **highly specialized food/resource requirement** (e.g., bamboo food requirements for Giant Panda), or 4) **highly specialized physical habitat requirements** (e.g., many of Hawaii's native forest birds are restricted to a very small range where specific moisture requirements are at their extreme tolerance level).

b) *Percent change* is calculated by the following formula:

$$\% \text{ change} = \frac{\text{final value} - \text{initial value}}{\text{initial value}} \times 100$$

$$\% \text{ change} = \frac{3.1 - 1.1}{1.1} \times 100 = 1.8 \times 100 = \textbf{180\% increase in density}$$

c) Other than population size and population density, *one additional factor* to consider in predicting the future dynamics of the population is (choices follow):

1. **Overall spatial arrangement:** clumped, uniform, or random. In nature, the pattern most often encountered is clumped—organisms arranging themselves according to the availability of resources needed to survive.
2. **Sex ratio:** a population's proportion of males to females. In monogamous species, a 1:1 sex ratio maximizes population growth, whereas an unbalanced ratio leaves many individuals of one sex without mates.
3. **Age structure:** the relative numbers of each age within a population. By combining with reproductive potential of the species, predictions can be made as to whether population will grow or shrink.
4. **Birth and death rates:** all preceding factors can influence the rates at which individuals within a population are born and die.

d) *Two* of the most likely factors causing the density trend in the last 9 years of the study are (choices follow):

1. continued effects of invasive species: preying upon nestlings or out-competing the native bird species for the limited food and habitat resources.
2. diseases, such as malaria or parasitic lice, causing the overall weakening and/or death of native bird species.
3. effects of global climate change on native forest vegetation, upon which the native birds depend for food and habitat. If temperature increases are causing native birds to seek higher elevation, they will be concentrated into a smaller area of the refuge, increasing competition for scarce resources needed for survival.

e) *Three* strengths of the experimental design for this field study are:

1. extensive *replication*: with a 21-year study there was more than one year of field data in order that long-term trends could be adequately assessed;
2. 15 transect lines were established throughout the refuge, with over 20 observation points along each transect. These *same transect lines and observation points were returned to each year* of the study; and
3. *standard observation times* (8 minutes at each observation point, no matter who the trained observer was).

# *Topic 3: Environmental Decision Making*

**1. (A) is correct.** Neoclassical, 20th-century economics assumed that all economic growth was good, resources were infinite, and that an anthropocentric (human-centered) worldview was prevalent. Internal costs and benefits are always considered, as these are the ones that can be estimated between the buyer and seller alone. Twenty-first-century economic policy is finally employing the view that all citizens deserve just and equal treatment, regardless of income, race, or ethnicity.

**2. (C) is correct.** The National Environmental Policy Act (NEPA) is the cornerstone of all U.S. environmental laws and requires that an environmental impact statement (EIS) be prepared for each project or action planned by the federal government.

**3. (B) is correct.** One who values all aspects of a functioning ecosystem (water cycles and air quality) is an ecocentrist. A biocentrist (A) values all living species, not just humans as anthropocentrists (C) would.

**4. (E) is correct.** Cap-and-trade systems are a kind of permit trading where an acceptable level of pollution has been determined by the government, which then issues permits to the various polluters/industries. An industry can receive credit for amounts it does not emit and can then sell this credit to other industries that are not able to meet the pollution limit.

**5. (A) is correct.** A cost-benefit analysis always includes internal costs, as these are the ones most easily estimated since they occur between the buyer and seller. External costs (B) are much more difficult to estimate (costs affecting a member of society other than the immediate buyer or seller) and are sometimes purposefully omitted in order to make the cost of a development project appear more environmentally friendly and economically feasible. The GDP (C) and GNP (D) have nothing to do with cost-benefit analyses. The intrinsic value (E) is the value ascribed to a thing/animal for its own sake and is most often not a dollar amount.

**6. (D) is correct.** Federal environmental legislation always takes precedence over state laws whenever they come into conflict on a specific issue.

**7. (C) is correct.** All "tragedy of the commons" scenarios involve a "commons" (a natural resource able to be utilized by everyone because of public "ownership"). The "tragedy" results from overutilization of the resource by the public, with everyone seeking to maximize their personal gain. With no limits to utilization, pollution and/or depletion of the resource results. Strip mining is not allowed in national parks, and (B), (D), and (E) involve *private* resources; (C) involves a *public* resource (the atmosphere) and the current pollution of this natural resource.

**8. (C) is correct.** The underlying goal of hydraulic fracturing (hydrofracking or fracking) is to increase supplies of natural gas, a fossil fuel less polluting than coal or oil. Natural gas is not a renewable energy source (E). The fracturing process is accomplished with deep drilling followed by high-pressure fluids (water, sand, chemicals) pumped into the fractures. Bubbles of natural gas trapped in the shale migrate through the fractures, gradually rising to the surface through the drilling pipe.

**9. (C) is correct.** External costs are those borne by someone not *directly* involved in an economic transaction. Examples include harm to citizens from water pollution, air pollution, and in this case, acid rain damage (C) caused by a nearby factory. All other answers represent internal costs: an employee's wages, energy operational costs, worker's compensation costs, and raw material extraction.

**10. (B) is correct.** The Montreal Protocol (1987) is considered the most successful effort to date in addressing a global environmental problem (stratospheric ozone depletion). CFCs have been dramatically reduced, and all major countries producing CFCs have ceased production. The Kyoto Protocol has not been as successful because many countries (including the United States) have not yet signed on. CITES has also had problems with enforcement. Choices (D) and (E) apply only to the United States and not international environmental law.

**11. (C) is correct.** The GPI (Genuine Progress Indicator) differentiates between desirable and undesirable economic activity and better indicates a society's well-being. GNP (Gross National Product) refers to the amount of goods and services purchased by households in a year, and GDP (Gross Domestic Product) includes only the goods and services produced within national borders; however, neither accounts for nonmarket benefits of factors such as volunteerism or external costs such as environmental degradation.

**12. (A) is correct.** Modern environmental policy aims to promote fairness among people and groups in the use of resources. Too many competing interests need to be addressed in order to preserve natural areas in the most pristine condition possible (D). Sustainable growth is preferred over promoting economic growth whenever possible (E). Natural resources are used for many purposes (aesthetic, ecosystem services, recreational), not just for economically important industrial products (B). Protecting the values of the landowner (C) is always balanced with competing needs, as sometimes a regulatory taking occurs when the government deprives landowners of some economic use of their private property.

**13. (B) is correct.** The 1997 study published in *Nature* estimating the economic value of ecosystem services was the work of Robert Costanza. Aldo Leopold (A) was a pioneering environmental philosopher promoting his "land ethic"; John Muir (C) was known for his preservationist ethic, was the founder of the Sierra Club, and was an early promoter of wilderness areas; Gifford Pinchot (D) was known for his conservationist ethic and was the first chief of the National Forest Service; and Rachel Carson (E) was the author of *Silent Spring*, which shed light on the hazards of DDT and other pesticides.

**14. (C) is correct**. The timber industry is regulated by the Department of Agriculture, since the U.S. Forest Service is under the authority of this federal department. The Department of the Interior has jurisdiction over the National Park Service and all endangered species. The Environmental Protection Agency is charged with setting standards for environmental quality, but does not have the authority to enforce. Enforcement is left up to the appropriate federal department.

**15. (D) is correct**. The graph showing methane concentration (mg CH₄/L) versus distance (m) to the nearest gas well shows that the highest concentrations of methane are found the closer one gets to the gas wells (the smaller distance); therefore, the relationship is an inverse one. The relationship of fracking to seismic activity is not depicted in this graph.

## Free-Response Question/Explanation

a) Conservationists could point to an **economic gain** with this federal decision because protecting this intact ecosystem would allow the entire population, not just the timber industry, to continue to realize and experience the value of the many ecosystem services. The value of the intact forest in preventing soil erosion, regulating oxygen/carbon dioxide and water cycles, storing and regulating water supplies in the region's watershed and aquifers, providing habitat for organisms to breed/feed/migrate, producing fish and game, and providing recreation such as ecotourism to one of the last remaining habitats of old-growth forest are just a few of the "hidden" **nonmarket values**. These ecosystem services are often not included in a cost-benefit analysis proposed by commercial interests because they are **external** to the immediate buyer-seller relationship.

b) The 1973 **Endangered Species Act**, prohibiting the government or private citizens from taking actions that would destroy a listed species or its habitat, is the central piece of legislation involved in this conflict. Even if no northern spotted owl is directly killed in the logging process, this act also protects the *habitat* of the species and requires that a *recovery plan* for the restoration of the species be instituted.

c) An **internal economic cost** is defined as one that is directly experienced by either the buyer or seller involved in the given transaction and is typically passed along to the consumer. In this case, the internal economic cost consists of the initial cost of all logging equipment, wages paid to the loggers, worker's compensation and insurances involved in the logging industry, and the fuel costs required for the continued operation and maintenance of the logging equipment. An **external economic cost** (externality) is one that affects other members (not only the buyer and the seller) of the society. The effects of water pollution (increased sediment) and air pollution (increased $CO_2$, $NO_x$, and $SO_x$) coming from this logging enterprise, as well as declines in desirable elements of the environment (fewer fish and wildlife), are the most obvious external economic costs.

**d) GDP** is the total monetary value of a nation's goods and services produced each year. GDP has been criticized as a progress indicator because economic growth is only indirectly related to quality of life. GDP *does not account for nonmarket values*, and it is not solely an expression of *desirable* economic activity. Instead, it includes all economic activity, so GDP can rise (and often does) when the economic activities driving it harm society and/or the environment. The **Genuine Progress Indicator (GPI)**, introduced in 1995, is a more realistic and accurate measure of progress because it takes into consideration positive contributions that are not paid for with money (volunteer work, parenting), then subtracts the cost of all negative impacts (crime, pollution, etc.) from the starting GDP. Although GDP has increased dramatically since 1950, per capita GPI has leveled off since 1975; therefore, one can conclude we are spending more money, but our lives are not much better in quality.

# *Topic 4: Human Population Dynamics*

**1. (C) is correct.** Multiplying the growth rate (which is a percentage) by the original population size results in the number of individuals that will be added to a population. When this resulting number is added to the original population, the final population size will result. Birth rates and immigration are always taken into consideration in determining the growth rate, but by themselves they cannot determine the exact number of individuals added to a population.

**2. (D) is correct.** A pyramid-shaped age structure diagram indicates that most of the population is pre-reproductive and that there is a significant momentum for continued and future population growth. The population is expanding and will continue to do so for more than one generation. A stable population diagram would generally have vertical sides, with equal numbers in each of the major age classes.

**3. (E) is correct.** The rapid growth of human population since the Industrial Revolution has been primarily due to technological advances in agriculture (the ability to raise more food) and improved medical care (reducing death rates and ultimately reducing birth rates as a result of lower infant mortality). Environmental resistance definitely applies to humans. Climatic change is only a very recent phenomenon, and although other species have been adversely affected, it has not affected human population size to date.

**4. (C) is correct.** Industrialization initiates the transitional stage of demographic transition. Declining death rates, which are due to increased food production and improved medical care, are the first significant change that occurs in a population. Birth rates decline *after* these falling death rates. Technological advances, although an important cause of increased food production and medical advances, do not immediately lead to population decline and improved status of the population. Although death rates and birth rates initially diverge, this change leads to a population increase because death rates decline first.

**5. (A) is correct.** The global growth rate is determined by considering only birth and death rates, since immigration and emigration are not factors on a global scale. The formula for global growth rate is $CBR - CDR \div 10$; $(15 - 12) \div 10 = 0.3\%$.

**6. (A) is correct.** Decreasing the average number of births per woman, the total fertility rate (TFR) is the most critical factor in controlling human population growth. The remaining choices, (B), (C), (D), and (E), are each important in controlling population growth; however, the TFR is the *most* critical and influential.

**7. (B) is correct.** With the world's fastest-growing economy over the past two decades, China is "demonstrating what happens when large numbers of poor people rapidly become more affluent" (Lester Brown). New material wealth and the resulting resource consumption are bringing unprecedented environmental challenges to China. Although population size continues to be large, China's actual growth rate has declined considerably (P). Technology has improved China's ability to grow more food and have improved medical care (T), but overall, (P) and (T) have not been as important as China's increasing affluence, resulting in increased resource consumption (A).

**8. (C) is correct.** The age structure diagram of Nigeria (b), with its very broad base, indicates this population is growing rapidly because there are large numbers of individuals in their reproductive years and even more in their pre-reproductive years, showing much future potential for growth. The age structure diagram of Canada is fairly stable (equal numbers in various age cohorts) and shows a population that is not aging rapidly. Age structure diagrams do not indicate anything about a country's progression through demographic transition.

**9. (A) is correct.** Nigeria, a developing country with most of its population below 40 years of age, will likely experience resource depletion and decreased quality of life due to a population that is growing rapidly and has great future momentum.

**10. (A) is correct.** The balanced age structure diagram of Canada suggests its total fertility rate (TFR) is less than the 2.1 replacement level value. Answers (C), (D), and (E) all indicate a high or increasing TFR, which the diagram does not indicate. With a zero TFR, there would be a very small to nonexistent base, which is not the case.

**11. (A) is correct.** The size of the U.S. population (P) is fairly insignificant compared to the great importance of affluence (A) and technology use (T). The addition of 1 American has as much environmental impact as the addition of 3.4 Chinese, 8 Indians, or 14 Afghans, all because individuals from affluent societies leave a considerably larger per capita ecological footprint.

**12. (E) is correct.** Replacement fertility is equal to 2.1 in stable populations. It takes two children to replace the mother and father, and the extra 0.1 accounts for the risk of a child dying before reaching reproductive age. If the TFR drops below 2.1, population size (in the absence of immigration) will shrink. With the growing populations in Africa, Latin America, and the Caribbean, the TFR is considerably greater than 2.1.

**13. (B) is correct.** The doubling time formula is: $DT = 70 \div \%$ growth rate. Transposing this formula, $\%$ growth rate $= 70 \div DT$. Hence, $\%$ growth rate $= 70 \div 30 = 2.3\%$.

**14. (D) is correct.** The U.S. age structure diagram has a small base, and most of its population is in their post-reproductive years, with the "baby boom" generation now approaching retirement age.

**15. (C) is correct.** For much of the 20th century, the growth rate of the human population rose from year to year. This rate peaked at 2.1% during the 1960s and has declined to 1.2% since then.

**16. (C) is correct.** The first major change to population dynamics during the demographic transition is that death rates decline while birth rates continue to remain high. This is known as the transitional stage, when the first significant change (or transition) occurs.

## *Free-Response Question/Explanation*

a) $\text{Percent change} = \dfrac{\text{Final value} - \text{Initial value}}{\text{Initial value}} \times 100$

In this case, $\% \text{ change} = \dfrac{(13.7 \times 10^8) - (5.0 \times 10^8)}{(5.0 \times 10^8)} \times 100 = \dfrac{8.7 \times 10^8}{5.0 \times 10^8} \times 100$

$\% \text{ change} = 1.74 \times 100 = \textbf{174 \%}$

b) China has experienced the first two stages of demographic transition.

In phase 1, **the pre-industrial stage**, high birth rates and death rates are characteristic and the country experiences very slow growth rates. This was during China's subsistence and predominantly agricultural years, up to approximately 40 years ago. Once industrialization and modern medical advances began to affect China's economy, phase 2, **the transitional stage**, began with the lowering of death rates followed by a very rapid growth in overall population size. It was at this time (in the 1970s) that China instituted the one-child policy in an attempt to reduce their very high TFR from 5.8, down to below replacement level (2.1). Economic incentives such as better access to schools, medical care, housing, and government jobs also were given to one-child families.

Most demographers hypothesize that China is currently in the early years of the third phase, industrialization. However, even though the growth rate for the country has dropped dramatically (0.5%), the country is still experiencing significant population growth (even a small percent of a very large number, 1.3 billion, results in significant growth!), adding millions to the population each year. Therefore, some demographers feel that China remains in stage 2, the transitional stage.

c) China's age structure base (0–15 years) is currently **more narrow** than the reproductive age classes (15–45) above it. The evidence for this is that the total fertility rate (TFR) in China is currently 1.5 children, significantly below the replacement level of 2.1, as a result of China's one-child policy reforms, which have resulted in many fewer children born today. The post-reproductive age classes farther above in the age structure diagram are pyramid-shaped and narrower than the reproductive age classes.

d) In order to determine the number of additional people that will join China's population, one must utilize the current growth rate (0.5%) and the current population size (~1.37 billion):

$$1,370,000,000 \times 0.005 = 6,850,000.00$$

$$1.37 \times 10^9 \times 5.0 \times 10^{-3} = \textbf{6.85} \times \textbf{10}^6 \textbf{ additional people}$$

e) Doubling time = $70 \div \%$ growth rate

$$70 \div 0.5 = \textbf{140 years}$$

# Topic 5: Earth's Land Resources and Use

**1. (C) is correct.** The B horizon, the subsoil, is located below the following layers: the O horizon (litter layer composed of decomposing organic material), the A horizon (topsoil), and the E horizon (leaching layer). The B horizon, located directly below the E horizon, accumulates the minerals and organic matter from the E horizon located directly above.

**2. (E) is correct.** No-till agriculture is the ultimate form of conservation tillage in which the land is left with a ground cover and crop residues after the harvesting process. To plant a subsequent crop, a "no-till drill" that cuts furrows through the O horizon of dead weeds, crop residue, and upper levels of the A horizon is employed. There is minimal disturbance of the soil.

**3. (B) is correct.** GM organisms involve the processes of recombinant DNA whereby genes of organisms as different as viruses and plants or bacteria and humans are combined in the same species. The genes for various desired traits exist in nature and are not artificially created in a lab (C); however, they can be isolated from the species where they exist and placed into species lacking this trait. The genes are not "modified and neutralized" (E), as then the effect of the desired trait would be lost.

**4. (D) is correct.** The initial application of the pesticide does not kill every individual, and the small number of survivors are the strongest/most pesticide resistant, having genes that allow for this better survival. When these survivors mate and pass their genes for resistance to their offspring, the next generation will have an even greater resistance to the pesticide. This is the natural selection process of evolution.

**5. (D) is correct.** Island biogeography explains how species come to be distributed in "habitat islands," and today these "islands" are best represented by national parks and other protected areas located in vast areas of development. Species richness, immigration, and extinction rates can be predicted from the "island's" size and its distance from the larger/main reservoir of species. Although Wilson and MacArthur's research originally took place on islands, the implications today for informed development of protected areas (particularly with regard to size and the adequacy of wildlife corridor locations) is the main application of their research.

**6. (B) is correct.** Second-growth forests are forests that grow back after the primary/virgin forest has been cut. Second-growth forests continue to grow in abundance worldwide and are not located mostly in British Columbia and Alaska, as this is the major location of primary/old-growth forests (E). Forests cannot be cut in wilderness areas; hence (C) is incorrect.

**7. (E) is correct.** Mining for coal and copper often occurs through strip mining, which removes surface layers of rock, exposing sulfide minerals that then react with oxygen and rainwater to produce sulfuric acid. As the sulfuric acid runs off, it can leach heavy metals from rocks, many of which are toxic to organisms. This process is referred to as acid drainage and is a significant water pollution by-product of strip-mining.

**8. (E) is correct.** Concern over declining crop diversity (because of the monoculture approach used by modern agribusiness) has led to the establishment of seedbanks, with the hope of preserving Earth's genetic assets for future agriculture.

**9. (D) is correct.** Overfertilization can easily lead to inorganic plant nutrients (primarily N and P) getting into waterways, where they stimulate algal growth, promote the eutrophication process, and result in hypoxic aquatic ecosystems. Adding too much fertilizer to cropland doesn't promote future fertile soils because the irrigation process and normal rainfall will usually leach these excess nutrients into waterways.

**10. (A) is correct.** As scientists have continued to study forest fragmentation, especially in the Amazon rainforest where forest studies have been going on since the early 1970s, they have found that small fragments lose more species, and lose them faster, than larger fragments. This is because smaller fragments contain fewer species, making these small areas more susceptible to outside threats because food chains are fewer, less complex, and less stable. Climate change has only exacerbated forest fragmentation issues, often eliminating higher elevation species that need cooler temperatures. Forests play a large role, not a minimal role, in the water, nitrogen, and carbon biogeochemical cycles.

**11. (C) is correct.** The Green Revolution began in the 1940s, when human populations appeared to be outgrowing their available food supply and their ability to produce food for this increasing population growth, especially in developing countries.

**12. (B) is correct.** Integrated pest management makes use of many varied practices used in combination. However, one of the central approaches involves using the natural enemy of a pest species to control the pest. IPM does not rely on methods of GM agribusiness, which are the focus of answers (C) and (D).

**13. (D) is correct.** Wilderness areas are the most protected federal land designation. No hunting, new road building, development, permanent structures, or harvesting of timber is permitted; however, low-impact recreation is allowed.

**14. (C) is correct.** The Healthy Forests Restoration Act encourages prescribed burning and was enacted in response to various large and destructive fires that occurred in the West after decades of a National Forest Service fire suppression policy. In implementing this policy, timber companies have mainly focused on the physical removal of small trees, underbrush, and dead trees.

**15. (D) is correct.** The Surface Mining Control and Reclamation Act (1977) mandates restoration efforts requiring companies to post bonds to cover reclamation costs before mining can be approved. This ensures that if the company fails to restore the land, the government will have the money to do so.

**16. (A) is correct.** IPM is the agricultural approach known as integrated pest management. It incorporates numerous *sustainable* farming techniques including biocontrol, use of limited chemicals when essential, crop rotation, and alternative tillage methods. Roundup Ready crops (D) are employed by modern agribusiness, where monocultures are the focus. CAFOs (C) refer to concentrated animal feeding operations, another modern agribusiness approach that is in contrast to sustainable agricultural techniques because of its negative effects on soils, water, and atmosphere.

**17. (D) is correct.** The process of leaching removes water-soluble minerals, in this case plant nutrients, from one location and transports them to another. The plant nutrients arising from the decomposition of organic matter in the O horizon can be leached out of the topsoil (where they would normally promote plant growth) under *un*sustainable agricultural practices such as over-irrigation.

**18. (B) is correct.** Humus is the dark, spongy, crumbly mass of material made up of complex organic compounds resulting from the partial decomposition of leaves, plant and animal remains, microbes, and their wastes. Soils with high humus content hold moisture well and are productive environments for plant growth.

**19. (D) is correct.** Since most plants cannot fix their own nitrogen, the addition of fertilizers is required to provide the plants with sufficient nitrogen for growth. GM crops that can now fix nitrogen into a usable form would require less fertilizer.

**20. (C) is correct.** Although the Green Revolution increased agricultural yields, the effects on the environment have not been as positive. The planting of crops in monoculture, large expanses of single crop types, has led to the need for increased pesticide use, as all plants in a field are genetically similar and equally susceptible to viral diseases, fungal pathogens, or insect pests. The Green Revolution did not employ methods of genetic engineering; thus, answers (D) and (E) are incorrect.

**21. (B) is correct.** The bar graph comparing various parameters of organic and conventional farming clearly shows that the energy input is lower with organic farming techniques (approximately 3,200 megajoules/acre/yr required for organic farming versus 4,500 megajoules/acre/yr required for conventional farming).

**22. (D) is correct.** Conventional farming techniques produce *more* greenhouse gas emissions, not less. Therefore, one needs to analyze the graphs quantitatively in order to determine exactly how much more greenhouse gases are generated. According to the data, conventional farming generates approximately 1,400 lb $CO_2$/acre/yr and organic farming generates approximately 900 lb $CO_2$/acre/yr. This is a 500-lb $CO_2$/acre/yr difference; 50% of 1,400 lb is a 700-lb difference, and is a larger than accurate difference.

**23. (E) is correct.** (Final − Initial) ÷ (Initial) × 100 = % change. Substituting in values from the graph, (170 − 140) ÷ (140) × 100 = 21%, with the answer (E) being the closest to this calculated value.

**24. (B) is correct.** By comparing the graph lines for industrialized and developing nations, one can see that the darkest line, representing developing nations, has the steepest slope, indicating that there is a greater rate of change (faster growth rate) occurring with GM crops in these countries. There is no information regarding world food supply provided by the graph.

**25. (A) is correct.** Conservation tillage is an array of approaches that reduce the amount of tilling relative to conventional farming, maximizing ground cover to protect soil erosion. One common definition is any method of limited tilling that leaves more than 30% of crop residue covering the soil after harvest. Although planting of some native grasses and polyculture approaches (the planting of multiple crops in a mixed arrangement) are both involved in some conservation tillage approaches, the remaining portions of options (B) and (D) are incorrect.

**26. (B) is correct.** Climate change scientists are concerned with the increasing levels of greenhouse gases in our atmosphere, primarily $CO_2$. Because of their involvement in photosynthesis, forests have the potential of consuming large amounts of $CO_2$, thereby sequestering (storing) carbon and removing it from the atmosphere. Although (A) and (C) are ecosystem services performed by forests, they are not as directly connected with climate change. Forests do not neutralize CFCs (D) or store methane (E), although both of these are greenhouse gases.

**27. (B) is correct.** Many environmental problems arise from feedlots; however, animal wastes moving into surface and groundwater during rain events is the most significant of the choices given. Animal wastes are rich in nitrates, leading to eutrophication of aquatic ecosystems when allowed to "escape" feedlots. Climate change concerns involve methane and nitrous oxide production from feedlots, not carbon dioxide (C). Use of antibiotics (D) *is* a feedlot issue; however, it involves a *human health problem* (increased antibiotic resistance of virulent microorganisms affecting human health).

**28. (E) is correct.** Sustainable forest certification involves following *all* steps in the life cycle of forest product production: from timber harvest to transport to pulping and the final production of paper or wood products. (C) is incorrect because all timber harvests today are from secondary growth harvests. A shelterwood approach (A) to forest management refers only to the harvesting of trees, and is not concerned with the entire "chain-of-custody" approach.

## Free-Response Question/Explanation

**a)** In the past, traditional agricultural techniques involved altering the gene pools of domesticated plants through selective breeding, preferentially mating individuals with favored traits (disease and pesticide resistance), so that offspring would inherit those traits. Plants would be chosen for breeding that seemed to offer natural resistance (showed little effects of pest damage) in order that those genes would be passed along to the next generation. However, the techniques that geneticists use to create genetically engineered organisms (GMOs) differ in several ways. First, selective breeding mixes genes from individuals of the *same or similar species*, whereas GMOs are created by *recombinant DNA processes,* which routinely mix genes of organisms as different as viruses and/or bacteria with plant crops. Second, selective breeding involves whole organisms living in the field, whereas genetic engineering works with genetic material (specific selected genes) in the lab. Third, traditional breeding selects genes that come together on their own, whereas genetic engineering creates new combinations in a controlled, lab environment.

**b)** i. Several notable examples of GMOs are (choices follow): Golden rice, Bt corn, Bt cotton, and Roundup Ready soybeans.

    ii. Golden rice has been engineered to produce health-enhancing beta-carotene to fight vitamin A deficiency in Asia and the developing world. Bt corn and Bt cotton have been engineered with genes from the bacteria *Bacillus thuringiensis*. The Bt genes kill insects, allowing less pesticide use, while increasing the yields of these plant crops. These GMOs can improve overall environmental q[ ] to less toxins/pesticides used.

iii. Roundup Ready soybeans have been engineered to tolerate herbicides, so that farmers can apply herbicides (Roundup) to kill the competing weeds without having to worry about killing their crops. An ecological concern is that herbicide-tolerant GM crops tend to result in *more* use of chemical herbicides, because farmers often need to apply increasing levels of herbicides each year, since these weeds are evolving resistance to herbicides.

c) In just over three decades, GM foods have gone from theory to big, industrial agricultural production. For example, worldwide, four of every five soybean plants are now transgenic, as are one of every three corn plants. Although GM crops are being grown by many countries, the promise of more easily produced crops on the Earth's marginal and poor land has not yet come to pass. This is mainly because the corporations that develop GM varieties cannot easily profit from selling GM seeds and herbicides to small farmers in developing countries, and this is often where population growth continues to surpass food production.

# *Topic 6: The Urban Environment*

**1. (D) is correct.** According to the graph of U.S. waste generation, one must look specifically at the per capita waste generation line and the appropriate scale, which is on the right side. In the year 2000, per capita waste generation was approximately 2 kg/person/day.

**2. (B) is correct.** The total waste generated in the U.S. in 1990 was approximately 200 million tons/year, and in 2010 it was 250 million tons/year. When calculating % change, one uses the following formula: Final value – Initial value ÷ Initial value × 100. Therefore, the calculation would be: $250 - 200 \div 200 \times 100 = 25\%$ .

**3. (C) is correct.** Smart growth is a city planning concept in which a community's growth is managed in ways that limit sprawl and maintain or improve residents' quality of life. The environment, economy, and community residents are all integrated into urban planning. By limiting sprawl, one is limiting impervious surfaces, fossil fuel consumption, and water pollution and increasing use of urban growth boundaries. Open space designation also increases, since urban areas are kept more concentrated.

**4. (A) is correct.** By definition, an $LD_{50}$ study is the lethal dose, or amount of toxicant, which will kill 50% of the population. Answers (C) and (D) produce percentages that are not 50% of the population. Answer (B) refers to the threshold effect.

**5. (B) is correct.** Since the late 1980s, recovery of materials for recycling has expanded, decreasing the pressure on landfills. U.S. waste managers are landfilling 53% of municipal solid waste, incinerating 13%, and recovering approximately 34% for composting and recycling.

**6. (E) is correct.** In 1980, the U.S. Congress passed the Comprehensive Environmental Response Compensation and Liability Act (CERCLA), which established a federal program to clean up U.S. sites polluted with hazardous waste from past activities. RCRA (the Resource Conservation and Recovery Act) is the federal legislation that specifies how to manage sanitary landfills to protect against environmental contamination. CERCLA now requires that industries monitor their hazardous waste output from "cradle-to-grave," and the past hazardous waste sites have been placed on a National Priority List for cleanup; however, these are not the names of this specific legislation. The EPA (Environmental Protection Agency) is the federal agency charged with conducting and evaluating research, monitoring environmental quality, and setting environmental standards.

**7. (A) is correct.** Urban sprawl involves increased use of automobiles and fossil fuel consumption. The burning of fossil fuel (gasoline) produces carbon dioxide and nitrogen-containing air pollutants, which lead to acid precipitation, tropospheric ozone, and urban smog, and potentially contributes to global climate change.

**8. (B) is correct.** Cities and towns are sinks for resources, since they must import from areas beyond their borders nearly everything they need to feed, clothe, and house inhabitants.

**9. (C) is correct.** Bisphenol A (BPA) is a chemical added to produce polycarbonate plastics, which are hard, clear types of plastic found in water bottles and many food containers. BPA has also been found to be an endocrine disruptor because it mimics the female sex hormone estrogen; that is, it is structurally similar to estrogen and can induce some of its effects in animals.

**10. (D) is correct.** All pesticides and synthetic chemicals are registered for use through the Environmental Protection Agency (EPA). The Food and Drug Administration (FDA) is charged with assessing the safety of all foods, drugs, and food additives.

**11. (D) is correct.** Paper products comprise the largest component (27%) of the municipal solid waste stream by weight, followed by food scraps (14.6%) and yard waste (13.5%). Heavy metals (9.1%) comprise a much smaller percentage, as does e-waste (1%), although the electronic waste category is now growing rapidly.

**12. (C) is correct.** The average American generates 2.0 kg (4.4 lb) of trash per day. This is considerably more than people in most other developed nations. Japan, for example, generates an average of 1.1 kg (2.4 lb). The 2013 estimate for the developing world was 0.55 kg (1.2 lb).

**13. (C) is correct.** Zoning is the practice of classifying areas for different types of development and land use.

**14. (D) is correct.** LEED certification is Leadership in Energy and Environmental Design, the leading set of standards for sustainable building construction. LEED certification does not require demolition, require that the construction take place in a greenway, or require that the structure be built entirely of recyclable materials. LEED buildings are often somewhat more expensive and do include designs that minimize water and energy consumption.

**15. (E) is correct.** Naturally occurring chemicals that are toxic are crude oil and radon gas. Carbon dioxide and nitrogen gas are both naturally occurring (B); however, they are not toxic. (C) and (D) both contain synthetic chemicals. (A) includes a living organism (*Salmonella*), not a chemical.

**16. (B) is correct.** Dose-response studies can test for *both* lethal effects ($LD_{50}$) and/or sublethal, effective dose effects ($ED_{50}$).

**17. (D) is correct.** By EPA definition, hazardous waste is waste that is one of the following: ignitable/flammable, corrosive, reactive/explosive, or toxic. Answer (D) contains three out of these four characteristics, making it a better answer than (C), which has only the toxic characteristic. Although *some* hazardous wastes (heavy metals and pesticides) can bioaccumulate, can biomagnify, and are fat soluble, these are not the EPA standard list characteristics.

**18. (A) is correct.** Reducing the amount of material entering the waste stream avoids the costs of disposal and recycling, helps conserve resources, minimizes pollution, and can save consumers and businesses money. To reduce waste, reuse and recycling [(B) and (C)] are often main strategies; however, they are only two of many strategies involved in reducing the overall waste stream. Composting (D) and incineration (E) also play a part in reducing the amount of material entering the waste stream. Recycling (C) has grown rapidly, and the EPA calls the growth of recycling "one of the best environmental success stories of the late-20th century."

**19. (B) is correct.** Asphalt pavement coverage, a widely used impervious surface found in U.S. watersheds where many roads are located, increases runoff into the watershed, as the precipitation can no longer be directly absorbed into the soil. As the upward sloping graph line indicates, a subsequent increase in chloride (salt) is produced in the environment. (E) is incorrect because *all* data points for suburban watersheds are below the dotted line labeled "chronic toxicity to freshwater life."

**20. (D) is correct.** The data indicates that 35% impervious surface coverage in a watershed will produce approximately 200 mg/L chloride. Answer (B) is too high, and (C) is too low. Answer (A) is characteristic of 250 mg/L and higher. Answer (D) is characteristic of *all* concentrations greater than 50 mg/L.

**21. (D) is correct.** Biomagnification is the concentration of toxicants in an organism caused by its consumption of other organisms in which toxicants have bioaccumulated. The higher the trophic level, the more severe the effects. Chemicals that are persistent (nonbiodegradable) and fat soluble are biomagnified. Of the choices given, PCBs are the only synthetic chemicals that have these characteristics.

**22. (E) is correct.** Synergistic effects are interactive effects between toxicants that result in an effect greater than or different from the effect predicted when each toxicant is experienced separately. These unpredictable effects are what make for challenging risk assessment studies, making it difficult to impossible to accurately assess the toxicant's risk.

**23. (A) is correct.** E-waste, or electronic waste, which is made up of discarded computers, cell phones, printers, hand-held devices, TVs, DVD players, and fax machines, contains heavy metals and toxic flame retardants. Most e-waste is being disposed of in landfills and incinerators, where it has traditionally been treated as conventional solid waste; however, recent research suggests that e-waste should instead be treated as hazardous waste.

**24. (E) is correct.** Incineration is the controlled burning process in which mixed garbage is burned at a very high temperature, along with metal-free waste. Incinerating waste reduces its weight by up to 75% and its volume by up to 90%. Many waste-to-energy (WTE) facilities use the heat produced by waste combustion to extract heat energy (D) and generate electricity (A); however, in the process of incineration, air pollution (dioxins, heavy metals, and PCBs) can enter the atmosphere. The life of a sanitary landfill is *increased,* not decreased (E), and this has been one of the perceived benefits of incineration, the great reduction of waste weight and volume.

### *Free-Response Question/Explanation*

a) The *annual* cost to dispose of the Jones household MSW:

$$4 \text{ persons} \times 4 \text{ lb/day} \times 365 \text{ days/yr} = 5{,}840 \text{ lb / yr}$$

$$5{,}840 \text{ lb} \times \frac{1 \text{ ton}}{2{,}000 \text{ lb}} = 2.92 \text{ tons} \times \$100/\text{ton} = \textbf{\$292/yr}$$

b) One specific problem that plastic presents *for the environment* is that it is durable and nonbiodegradable. Because plastics do not decompose, they take up significant volume in landfills, causing landfills to reach their capacity much more quickly. Plastics also pose a significant environmental hazard for wildlife. Some plastics (bags and other packaging) are mistakenly consumed as food by wildlife, leading to digestion obstruction and often death. A problem *for human health* resulting from plastics is their role as potential endocrine disruptors. These are chemicals that interfere with the normal functioning of hormones in an animal's body. Males normally make estrogen and then convert it to testosterone; however, endocrine-disrupting hormones are thought to interfere with this process, resulting in higher concentrations of estrogen in males, producing feminization characteristics and reproductive abnormalities. Bisphenol A is one of many plastic chemicals that appear to mimic the female hormone estrogen. Many plastic products also contain another class of hormone-disrupting chemicals called phthalates. Health research on phthalates has linked them to birth defects, breast cancer, reduced sperm counts, and other abnormal reproductive effects.

c) In toxicology, the standard method of testing with lab animals is by setting up a dose-response experiment. The strength of a chemical's effect is determined by the number of animals affected at different specific doses. Once the dose-response data are plotted, toxicologists can calculate the specific toxicity by determining the $LD_{50}$ value (lethal dose-50%), which is the amount of toxicant it takes to kill 50% of the test organism population. A high $LD_{50}$ indicates low toxicity and a low $LD_{50}$ indicates high toxicity.

d) *LEED certification* (Leadership in Energy and Environmental Design) is the leading set of standards for sustainable building. This certification process focuses primarily on assessing a building's minimal use of water and energy, building with sustainable materials, and recycling a building's wastes. Since the Jones home is a historic home, already built, they could consider retrofitting with low-flush toilets and low-flow showerheads (minimizing water use), increasing insulation, and installing thermal windows and compact fluorescent bulbs throughout their home (minimizing energy usage).

# Topic 7: Water Resources and Pollution

**1. (E) is correct.** Human wastewater/sewage is primarily organic material that will require significant oxygen in order to be decomposed by the bacteria and other microorganisms living in this body of water. This will produce a high BOD, or biochemical oxygen demand. Fecal coliform are bacteria that are indicator species for potentially harmful microorganisms associated with sewage contamination, since they are normally found in the mammalian and bird (warm-blooded) intestinal tracts. (B) is incorrect because human sewage contamination would not produce low BOD readings, and the fecal coliform presence could be due to the local wild bird and mammal presence.

**2. (B) is correct.** Non-point source pollutants are diffuse sources, often consisting of many small, spread-out sources such as a residential lawn, golf course, and feedlot. Sediment loading of a tributary (smaller river) would also come from a wide area of land through which the tributary flowed. The effluent from a sewage treatment plant represents a point-source pollutant, as it is a discrete, specific spot where the sewage is discharged through a pipe.

**3. (E) is correct.** Ocean acidification is the process by which today's oceans are becoming more acidic as a result of increased carbon dioxide concentrations in the atmosphere. Ocean water absorbs $CO_2$ from the air, forming carbonic acid, which dissociates into bicarbonate ions and hydrogen ions. The hydrogen ions then combine with carbonate ions present in seawater. It is these carbonate ions that are essential in the building of calcium carbonate shells of marine organisms and that are being removed by the ocean acidification process. Also, the lowering of the ocean's pH as a result of the presence of carbonic acid is further dissolving the already formed shells of many marine species.

**4. (D) is correct.** Oysters are filter-feeding marine invertebrates that have been the foundation of one of the most economically valuable fisheries in the Chesapeake Bay estuary for centuries. In the last several decades, oysters have also been overharvested and the nutrients they would normally be removing from the water column have remained, stimulating algal blooms and hypoxia. Kelp (A), otters (B), and cod (C) do not occur in the Chesapeake Bay estuary, nor are there offshore drilling platforms located off the Middle Atlantic states.

**5. (D) is correct.** A eutrophic pond is one that has high nutrient and low oxygen conditions. These conditions are most often caused by excess nutrients (possibly fertilizer runoff into the pond) stimulating phytoplankton and algal growth of this aquatic system. Endemic (native) algae in this pond could definitely be affected by excess nutrients, and this would make the most realistic parameter to study, given the remote inland location (no salinization effects) and no industrial effects (heavy metals). Pesticide application is not allowed in wilderness areas (C), and aquatic invertebrates do not directly affect native mammal populations, which are terrestrial (E).

**6. (A) is correct.** Secondary wastewater treatment involves *biological* means to remove the remaining contaminants after primary treatment (a physical/mechanical removal of materials) has taken place. Aeration basins provide oxygen to the aerobic microorganisms/bacteria that are decomposing the remaining organic wastes. When secondary treatment is completed, often 90% of the organic wastes have been removed through this decomposition process.

**7. (E) is correct.** Record-low levels of precipitation coupled with record-high heat have kept California and the American Southwest in drought conditions since 2012. According to a 2015 NASA study, the accumulated precipitation/snowpack deficit in the Sierra Nevada mountains (the source of most of California's water) was 20 inches—the equivalent of an entire year's average rainfall. The El Niño winter of 2015–2016 provided California with only a slight reprieve from the drought in northern and eastern California, but southern California ended the year with lower than average levels of rainfall.

**8. (D) is correct.** Trees are a natural archive of climate data because they grow at varying rates, depending on moisture, temperature, and other climate factors. Using data from tens of thousands of samples of tree rings from the U.S., Canada, and Mexico has provided an archive of past climates, which is now being used to predict future droughts and climates.

**9. (E) is correct.** Aquifers, underground water reservoirs, have regularly been polluted by industry and agriculture. Most pollution control efforts have focused on surface waters, as groundwater is hidden from view and difficult to monitor. Since agriculture (70%) and industry (20%) represent much more significant water-use sectors than residential (10%), their contribution to hazardous waste disposal (heavy metals, pesticides, volatile organic compounds) is by far the most significant contribution to aquifer contamination. Methane and carbon dioxide gases are not injected/pumped into aquifers (C), but are released into the atmosphere.

**10. (C) is correct.** An oligotrophic pond is one that has low nutrient and high oxygen conditions. In order to maintain these conditions, you would want to *avoid* inputs of nutrients (organic wastes or fertilizers/nutrients). Runoff from an athletic field would most probably contain fertilizers from the lawn-care provider; therefore, the possibility of stimulating algal growth leading to eutrophic conditions would be enhanced. Decreasing the calcium available to bottom-dwelling organisms is a sign of ocean acidification (not in a freshwater pond). Answers (A), (D), and (E) are opposite to what one would want to pursue to maintain oligotrophic conditions.

**11. (B) is correct**. Bycatch is the unintentional catch of nontarget species in fishing. It can include bottom-dwelling invertebrates (shellfish species), noncommercial pelagic fish, marine mammals (dolphins and smaller whale species), and sea turtles.

**12. (D) is correct**. Nitrogen flux is read along the right-hand side of the graph. After determining that 1993 was the year in which the nitrogen flux was the highest (following the graph's solid black line), one can then read this value as slightly over 200,000 metric tons, approximately 210,000.

**13. (B) is correct.** The Gulf of Mexico's waters would be considered an environment able to support a high diversity of fish and invertebrates if dissolved oxygen is not a limiting factor. In hypoxic waters (<2 mg/L dissolved oxygen), very few forms of aquatic life are able to survive. Therefore, one needs to find the year in which the bar on the graph is the *lowest* (2000), as the bars indicate the hypoxic area in square kilometers.

**14. (C) is correct.** Determining the dissolved oxygen (DO) content of a body of water will immediately inform one of the possibility of eutrophication processes. When a lake becomes eutrophic, microbial decomposition of the organic matter results in low oxygen (hypoxic) conditions. Measuring DO will not provide any *direct* information regarding heavy metals (E), bacterial parasitism of the aquatic invertebrates (D), acid precipitation (A), or aquatic biodiversity.

**15. (D) is correct.** Red tides are a harmful algal bloom consisting of algae that produce reddish pigments that discolor coastal and open ocean waters. Like all algal blooms, the algae are stimulated to bloom with increased nutrient inputs (fertilizer and feedlot runoff into fresh water that subsequently flows into coastal waters).

**16. (C) is correct.** Surface ocean waters are at the surface because they are less salty and warmer, both physical conditions that decrease the density of a body of water. With increased salinity and/or decreased temperature, ocean water becomes denser and will sink.

**17. (C) is correct.** Agricultural irrigation represents 70% of all water used globally. Industry accounts for roughly 20%, and residential and municipal uses for only 10%.

### Free-Response Question/Explanation

**a)** Reasons why the advent of industrial trawling would produce the dramatic *increase* in cod harvested are multifaceted, but are focused around the increased pressure exerted on the fish stocks resulting from (choices follow):

1) New fishing gear/technologies used to capture fish in great volumes (driftnets, longline fishing, and bottom trawling) are highly efficient in finding and harvesting marine fish.
2) The use of fossil fuels and large factory ships allow the harvesting and subsequent processing/freezing/storing of large volumes of fish before the need to return to shore to offload the harvest.
3) Faster factory ships allow for making return trips to fishing grounds more quickly with fewer hours/days lost to nonfishing activities at port.

**b)** Based on the data provided by the graph, the 10-year period during which the greatest decline in cod fishing took place was from **1970 to 1980**. This occurred approximately 10 years after the advent of industrial trawling. The rate of decline, or rate of change, is determined by the following formula:

$$\text{Rate of decline} = \text{Initial} - \text{Final} \div \text{Time Interval}$$

$$(800,000 \, \text{tons}) - (140,000 \, \text{tons}) \div 10 \, \text{years}$$

$$= 666,000 \, \text{tons} / 10 = \textbf{66,000 tons per year}$$

**c)** The commercial fishing practice of bottom trawling involves dragging a weighted net along the ocean floor to catch groundfish (cod), which are benthic species. Driftnetting uses long, transparent nylon nets arrayed to drift with currents so as to capture passing fish. These are held vertical by floats at the top and weights at the bottom of nets, and they hang from the ocean surface into pelagic waters. These two fishing methods differ in the deployment location of the nets, and therefore, the targeted species. Driftnetting usually targets schooling species that traverse open water such as bluefin tuna, swordfish, and squid, whereas bottom trawling targets groundfish, such as cod. Both of these methods are similar in their significant impact on bycatch; that is, both of these industrialized fishing methods involve the harvesting of many nontarget species. They differ in that bottom trawling is known to be extremely damaging to ocean bottom habitat. Currently, driftnet fishing is illegal because up to 80% of harvested species can be bycatch.

**d)** The main categories of marine pollutants are oil, plastic debris, mercury/heavy metals, and excess nutrients. Because this area off the northwest coast of Canada is not a prime agricultural area or offshore oil drilling location, it is less likely that excess nutrient (from agriculture) or oil seepage from offshore drilling would be major pollutant categories. However, because this area experiences such high fishing pressure, plastic, which comes from lost or intentionally discarded nets, could play a large role in harming marine life. Also, since all the world's oceans are interconnected, plastic debris can now be found everywhere, even in the most remote of ocean locations. Because plastic is designed not to degrade, it can drift for decades before washing up on beaches.

In addition, with the use of fossil fuels increasing globally, many of the fastest growing nations are burning coal in order to fuel their industrialization. Mercury and other heavy metals are emitted into Earth's atmosphere during the combustion of coal. After settling onto land and water, mercury bioaccumulates in animal tissues (including invertebrates and fish) and then biomagnifies as it makes its way up the marine food chain. As a result, fish at high trophic levels can contain substantial levels of mercury; hence, it is probable that many fish and shellfish species inhabiting the Grand Banks have experienced mercury contamination.

# *Topic 8: Atmospheric Science*

**1. (A) is correct.** Worldwide, average sea levels have risen 24.1 cm (9.5 in.) in the past 135 years of monitoring, reaching an average rate of 3.4 mm/yr from 1993 to 2016. These numbers represent a vertical rise in water level, and on most coastlines a vertical rise of even a few inches translates into a great many feet of water movement inland.

**2. (E) is correct.** Primary pollutants can be changed into secondary pollutants by the interaction of sunlight and water with the primary pollutant. All primary pollutants are those that come directly out of a smokestack, exhaust pipe, or other natural emission source. Secondary pollutants have all undergone chemical change. When $SO_2$ (primary pollutant) reacts with water vapor in the atmosphere, the secondary pollutant $H_2SO_4$ (sulfuric acid) is produced.

**3. (C) is correct.** In May 2013, scientists reported that tropospheric $CO_2$ levels had reached the 400 ppm level for the first time in 800,000 years. These levels have continued to rise and are currently *slightly* over 400 ppm, but are not 415 ppm.

**4. (E) is correct.** Methane concentrations have increased approximately 150% since 1750, primarily because of livestock metabolic wastes (more feedlots), disposing of more organic matter in landfills, and growing more crops, such as rice, to feed the increasing human population growth. Although carbon dioxide has also increased, from 280 ppm to over 400 ppm, this change is *much less* than a twofold increase.

**5. (C) is correct.** All of the responses are primary pollutants, coming directly from a factory's smokestack, except ozone $(O_3)$. This is a secondary pollutant that forms in the troposphere as a result of the interaction of sunlight, heat, nitrogen oxides, and volatile carbon-containing chemicals.

**6. (C) is correct.** Natural, unpolluted rainfall is slightly acidic as a result of the chemical interaction between the carbon dioxide and water vapor normally occurring in the atmosphere. This reaction produces carbonic acid, a weak acid that raises the pH of most unpolluted rainfall to 5.6.

**7. (B) is correct.** The largest source of U.S. emissions is electricity generation, accounting for 40% of the U.S. carbon dioxide. Transportation is the second-largest source of greenhouse gas emissions. This is easily understood when one recognizes the lack of mass transit systems in most areas of the U.S. The much larger-than-average size of automobiles driven by many Americans, as compared to Europeans and developing Asian countries, is also a contributing factor to high fuel usage and increased emissions from transportation.

**8. (D) is correct.** The Kyoto Protocol mandated signatory nations, by the period 2008–2012, to reduce emissions of six greenhouse gases to levels below those of 1990. Although the United States still has not signed this international treaty, over 100 developed and developing nations have ratified it and have begun to decrease their emissions while still maintaining their standard of living and continuing to be competitive on the international world market.

**9. (A) is correct.** The six "criteria pollutants" are $NO_2$, $SO_2$, particulates, CO, $O_3$, and lead (Pb). Answer (A) includes these first three. Answers (B) and (C) include $CO_2$, which is *not* a criteria pollutant monitored by the EPA, nor is nitrous oxide (NO), answers (D) and (E).

**10. (C) is correct.** The EPA regulates the six "criteria pollutants" that are judged to pose especially great threats to human health. These are $CO, SO_2, NO_2$, tropospheric ozone $(O_3)$, particulate matter, and lead. Carbon dioxide, although an emission of great concern because of its role in global climate change, is not currently regulated by the EPA.

**11. (C) is correct.** Acid deposition has many impacts; however, one of the most significant is that the acids leach nutrients such as calcium, magnesium, and potassium ions from the topsoil, altering soil chemistry and harming plants and soil organisms. Acid precipitation also "mobilizes" toxic metal ions, such as aluminum, zinc, mercury, and copper, changing them from insoluble to soluble forms. The presence of these ions in the soil does not make the soil more productive (A), but actually hinders water and nutrient uptake by plants.

**12. (D) is correct.** The greenhouse effect is due to certain gases of Earth's atmosphere (water vapor, ozone, carbon dioxide, nitrous oxide, and methane) absorbing the infrared radiation emitted from the surface of Earth. After absorbing this radiation, these gases subsequently re-emit this energy in all directions, with some being lost to space. However, some of this radiation travels back downward, warming Earth's atmosphere, specifically the lowest layer, the troposphere.

**13. (D) is correct.** The Montreal Protocol is the international treaty ratified in 1987 in which almost 200 signatory nations agreed to restrict production of chlorofluorocarbons (CFCs) in order to forestall stratospheric ozone depletion. This international treaty is considered the most successful effort to date in addressing a global environmental problem.

**14. (B) is correct.** The IPCC is an international panel, established in 1988, and consists of hundreds of scientists and government officials. It has released numerous reports summarizing thousands of scientific studies, addressing the impacts of current and future climate change. These reports have focused on measuring global and regional physical indicators (such as temperature and precipitation), social indicators (sea level rise), and biological indicators (studies of species range shifts). Oceanic photosynthesis and global productivity have not yet been specifically studied for these reports.

### *Free-Response Question/Explanation*

**a)** Three sources of data currently being used to study and document global climate change are (choices follow):

1) **Ice core data**—By examining the trapped air bubbles in ice cores, scientists can determine the atmospheric composition, greenhouse gas concentration, temperature trends, snowfall, and even frequency of forest fires and volcanic eruptions at the time these bubbles of gas were trapped.

2) **Sediment core data** beneath bodies of water—Sediments often preserve pollen grains from plants that grew in the past. Because climate influences the types of plants that grow in an area, knowing what plants were present can tell one a great deal about the past climate at that place and time.

3) **Tree ring data**—By examining the width of each tree ring, scientists can determine the amount of precipitation available for plant growth.

4) **Coral reef biochemistry**—Since living corals take in trace elements from ocean water as they grow and deposit their calcium carbonate skeletons, their growth bands give clues to ocean conditions at the time these $CaCO_3$ bands were deposited in the coral skeleton.

Each of the above paleoclimate data sources can then be compared to direct measurements of current climate and atmospheric gas concentrations. For example, direct measurements of $CO_2$ concentrations began in 1958 at the Mauna Loa Observatory in Hawaii and show an increase from 315 ppm in 1959 to over 400 ppm today. This atmospheric $CO_2$ concentration is now at its highest level by far in over 800,000 years.

b) The specific trends the IPCC 2014 report documents have been categorized according to physical, social/human, and biological/ecosystem effects. They have been further categorized as already observed consequences and future predicted trends and impacts. This question is asking for an explanation of the *already observed consequences/impacts* in the above categories.

Global or regional **physical effects already observed** are (choices follow):

1) Earth's *average* surface temperature increased 1.1°C in the past 100 years.
2) Sea level rose by an *average* of 24.1 cm (9.5 in.) in the 20th century.
3) Ocean water became more acidic by about 0.1 pH unit in the past 100 years.
4) Storm surges have increased over the past 100 years.

**Human/social effects** already observed are (choices follow):

1) Farmers and foresters have had to adapt to altered growing seasons.
2) Poorer nations and communities are suffering more from climate change because they rely more on climate-sensitive resources and have less capacity to adapt.

**Biological/ecosystem** effects already observed are (choices follow):

1) Species ranges are shifting toward the poles and upward in elevation.
2) The timing of seasonal phenomena such as migration and breeding is shifting.

c) Achieving **carbon neutrality** is the condition in which no net carbon is emitted by the individual (in this case), business, government, university, or utility. Since electricity generation is the largest source (40%) of U.S. $CO_2$ emissions, and transportation is the second-largest source, changing human behaviors that will impact electricity usage and transportation are the most important first approaches to take. Specific actions that reduce residential electrical consumption include using natural lighting during the day by turning off all lights, using only essential electrical appliances, using natural cooling by opening windows rather than air-conditioning, and completely turning off electrical appliances to avoid phantom power usage when an appliance is on "standby." Using only newer energy-efficient technologies such as the EPA's Energy Star–rated appliances (Energy Star washing machines and refrigerators) can cut one's $CO_2$ emissions by as much as 440 lb/appliance annually, and replacing all light bulbs with compact fluorescent bulbs can reduce energy use for lighting by 40%. The single most effective action to take regarding transportation would be to choose human-powered transportation methods such as walking or bicycling. The next most effective approach would be to rely on public transportation and to attempt to completely phase-out whenever possible the conventional automobile, which is very inefficient (only 13–14% of the energy from a tank of gas actually moves the typical car down the road).

**d)** To calculate the increase in sea level, ***in meters***, over the next 50 years:

$$3.4 \text{ mm/year} \times 50 \text{ years} = 170 \text{ mm}$$

$$170 \text{ mm} \times \frac{1 \text{ meter}}{1,000 \text{ mm}} = \textbf{0.177 meters over the next 50 years}$$

# *Topic 9: Energy Resources and Impacts*

**1. (C) is correct.** The first law of thermodynamics (B) states that the overall *quantity* of energy is conserved in any conversion of energy (converting the potential energy of coal into the kinetic energy of moving atoms and heat). However, the second law states that the *quality* of energy will change from a more-ordered state to a less-ordered state. That is, systems tend to move toward increasing entropy. During combustion, the highly organized and structurally complex coal or oil transforms into a residue of carbon ash, smoke, and gases such as $CO_2$ and $H_2O$ vapor, as well as producing heat. Some of this heat escapes without being utilized to make steam or turn the turbine in an electrical power plant.

**2. (D) is correct.** "Peak oil" is the term used to describe the point of maximum production of petroleum in the world, after which oil production declines. This is also expected to be at roughly the halfway point of extraction of the world's oil supplies.

**3. (E) is correct.** With fossil fuels producing approximately 82% of U.S. energy consumption, there is no other fuel source (biomass, hydroelectric, nuclear) that comes close to competing. Electricity generation utilizes *more* energy than the transportation sector (A); nuclear energy accounts for approximately 10.6% of electricity generation, in contrast to coal's 41.3% (C); and biofuels (D) have a very low EROI (approximately 1.3:1).

**4. (D) is correct.** The long-term, safe storage of the radioactive wastes produced by all nuclear power plants remains the most serious *unsolved* dilemma surrounding nuclear power today. Although nuclear power poses small risks of large accidents, as have been demonstrated by Three Mile Island, Chernobyl, and the recent Fukushima, Japan, accident, it is the permanent waste disposal issue that has not yet been solved by any nation.

**5. (E) is correct.** Passive solar energy is based on a building's design, orientation to the sun, and the use of construction materials that can insulate and store heat. South-facing windows are necessary to maximize the absorption of sunlight in winter, and stone floors add to the thermal mass required to store the sun's radiant heat for release at night. When evergreens are planted outside south-facing windows (D), the sunlight is blocked from entering, and when deciduous trees are planted outside north-facing windows (C), they have no buffering effect in the winter because they have no leaves.

**6. (C) is correct.** Geothermal energy is one form of renewable energy that does not originate from the sun. Instead, geothermal energy is thermal energy that arises from beneath Earth's surface due to the radioactive decay of elements amid high pressures deep in the planet's interior.

**7. (C) is correct.** Oil represents 35% of U.S. energy consumption, a significantly greater portion of our energy budget than coal (18%), natural gas (28%), or nuclear (8.5%) fuels.

**8. (D) is correct.** When all fossil fuels are burned, their carbon atoms chemically react with oxygen atoms, producing $CO_2$ as a waste product from this combustion reaction. $CO_2$ is the most prevalent greenhouse gas, and its atmospheric concentration has been increasing significantly since 1958 when Charles Keeling began collecting detailed $CO_2$ data at the Mauna Loa Observatory in Hawaii. Carbon is not liberated back into the carbon cycle (A); however, carbon in the form of $CO_2$ is returned to the carbon cycle. There are current studies focusing on whether increased $CO_2$ concentrations will increase plant growth. Normal "unpolluted" rainfall is naturally slightly acidic, with a pH of 5.6, because of the carbonic acid resulting from the reaction of atmospheric $CO_2$ with $H_2O$. However, it is the $NO_x$ and $SO_2$ atmospheric pollutants that significantly lower the pH of rainfall *below 5.6*, qualifying it as acid rain.

**9. (B) is correct.** Worldwide, approximately 41% of electricity is generated by coal; natural gas comes in far behind, at approximately 28%. Nuclear energy accounts for only 11% (A). Biofuels and biomass produce $CO_2$ emissions, even though this $CO_2$ is considered *modern* carbon, in contrast to the carbon sequestered for millions of years in fossil fuels (C). Nuclear fuel produces *no* $CO_2$ emissions. Currently, hydroelectric power represents the *highest* EROI (energy returned on investment) of any renewable or nonrenewable energy source, at 84:1 or more (E).

**10. (C) is correct.** Worldwide, the most widely used renewable energy resource is biomass, representing approximately 10% of the total world energy production. Nuclear energy (E) is a *non-renewable energy source*. Hydroelectric comes in second, with only 2.4% of the market (B). Wind and solar together share less than 1% of the world energy production.

**11. (D) is correct.** Photovoltaic (PV) cells convert sunlight to electrical energy by making use of the photovoltaic effect. This involves *sunlight* striking one of a pair of plates made primarily of silicon, a semiconductor. The light causes the silicon to release electrons that are attracted by electrostatic forces to the opposing plate, generating electricity.

**12. (D) is correct.** Nonrenewable fossil fuels provide approximately 81% of the world's energy resources. Concerns over the finite and shrinking quantity of all fossil fuels and their negative environmental impacts, especially with regard to air pollution emissions of $CO_2$ and the climate change implications, have motivated the development of solar, geothermal, and wind energy technologies.

**13. (E) is correct.** EROI stands for energy returned on investment and is a ratio that is useful when assessing energy sources. EROI = Energy returned/Energy invested. Higher ratios mean that we receive more energy from each unit of energy than we invest in the exploration, extraction, and transportation costs for the energy resource to get to the market or utilization site.

**14. (D) is correct.** "Clean coal" technology includes removing $SO_2$ with scrubbers (A), capturing and sequestering carbon emissions ($CO_2$) into rock formations deep underground (B and E), and converting coal to *syngas* through gasification (C). Carbon *cannot* be removed from coal *before* combustion, and such removal is not a part of any clean coal technology or approach currently being researched.

**15. (A) is correct.** Fossil fuels (~81%), biomass (10%), nuclear (4.8%), hydroelectric (2.4%), and new renewables (1.2%) represent the current best estimate for the world's total energy consumption by source.

**16. (B) is correct.** All biofuels (ethanol and biodiesel) are liquid fuels from biomass sources. All biomass consists of organic material derived from living or recently living organisms and contains chemical energy that originated ultimately with sunlight and photosynthesis.

**17. (B) is correct.** Over the past several decades, the fastest-growing source of electricity generation using new renewables has been wind power, doubling approximately every 3 years. Denmark is the global leader, supplying 40% of its total electricity needs from its wind farms.

**18. (C) is correct.** The Keystone XL Pipeline project refers to the Canadian tar sands mining occurring in northern Alberta. The tar sands slurry recovered from the massive open pit mining is being transported via a trans-Canadian pipeline 2,200 miles through the central United States to ports in Oklahoma and Illinois, and on to Gulf Coast refineries.

**19. (D) is correct.** Coal continues to be the main fuel used to generate electricity in the United States primarily because we have such large reserves of this fossil fuel. Coal provides half the electrical generating capacity of the United States and is currently powering China's surging economy.

**20. (C) is correct.** A feed-in tariff system is the practice whereby utilities are mandated to buy extra power from anyone who can generate it from a renewable energy source and feed it into the electric grid. Net metering (A) is the process by which owners of houses with PV systems or wind turbines can sell their excess solar or wind energy to their local power utility. Whereas feed-in tariffs award producers with prices above market rates, net metering offers market-rate prices.

### Free-Response Question/Explanation

a) The average growth rate is calculated using the following formula:

$$\frac{\text{Final value} - \text{Initial value}}{\text{Initial value}} \times 100 = \% \text{ Change (Growth rate)}$$

For wind energy : $\dfrac{182,000 - 120,000}{120,000} \times 100 = \sim \textbf{51.7\% growth}$

For solar : $\dfrac{33,000 - 7,300}{7,300} \times 100 = \sim \textbf{352\% growth}$

b) In the past several years, solar energy (PV and CSP combined) is growing much faster than wind energy in the U.S. (approximately 300% faster according to the above 2011–2014 data). One reason for this difference in growth is that solar heating systems do not need to be connected to an electrical grid and are thus more useful in isolated locations. This important feature shows that solar energy need not be confined to only affluent suburban communities. A second reason for the difference is that results of recent research have demonstrated a growing concern for noise pollution and wildlife deaths (especially migrating birds and bats) associated with wind turbines. Over the past four decades, solar, wind, and geothermal energy sources have grown far faster than has the overall energy supply. Although the long-term leader in growth over this period has been wind power (growing by nearly 50% each year since the 1970s), *in recent years* (as the above data show), solar power has grown far faster than wind power. Because solar and wind energy both started from such low levels of use, it will take many years for them to potentially catch up with conventional energy sources.

c) The goal of any feed-in tariff system is to promote the use of the new technology with economic incentives. Utilities are mandated to buy power from anyone who can generate extra power (over their needs) from renewable energy sources and then feed it back into the electric grid. Under this system, utilities must pay guaranteed premium prices for this power under a long-term contract. As a result, homeowners in Washington have moved to install more and more PV panels each year and they are selling their extra solar power to the utilities at a profit, as feed-in tariff systems significantly award producers with prices *above* market rates. It is assumed that the $0.15/Kwh electricity rate is above the current cost of electricity in Washington. Under this incentive, to further encourage local "green collar" industry in Washington, a rate of $0.54/Kwh (for extra electricity generation that can be fed into the grid) will be credited to homeowners using PV solar cells *manufactured in Washington.*

**d)** The major environmental advantage of using wind energy is that wind power produces no emissions once the equipment (tower and wind turbine) is manufactured and installed. The EPA has estimated that the amount of $CO_2$ pollution that all U.S. wind turbines together prevent from entering the atmosphere is greater than the emissions from 14 million cars each day. Overall, there are no air or water pollution issues associated with wind power. Secondly, wind is a nondepletable resource—the amount of wind available tomorrow does not depend on how much we use today. Wind energy is a clean and free resource.

However, the main reasons why wind power continues to represent less than 1% of current U.S. energy sources are 1) wind is an intermittent resource over which we have no direct control when it will occur; 2) wind also varies greatly with geographic location (most of North America's people live near the coasts, far from the Great Plains and mountain regions that have the best wind conditions); and 3) although there seems to be wide public approval of existing wind projects and the concept of wind power, newly proposed wind projects have often elicited the not-in-my-backyard (NIMBY) syndrome among people living nearby. Also at play is the lack of factoring in the external costs associated with fossil fuel use (air pollution, habitat destruction associated with mining, and human health issues associated with poor air quality), factors that seem to make conventional energy sources more economically accessible.

# Part IV

## *Practice Test*

On the following pages is a practice test that approximates the actual AP Environmental Science Examination in format, types of questions, and content. Set aside three hours to take the test. To best prepare yourself for the actual AP exam conditions, use only the allowed time for Section I and Section II. **No calculators may be used** in Section I or Section II of the examination.

# Environmental Science

## *Section I*

**Time—1 hour and 30 minutes**

**Directions:** This section of the examination contains 100 multiple-choice questions. Each question is followed by five suggested answers. Select the one lettered choice that *best* answers each question or best fits the statement. **No calculators may be used on the examination**.

**Questions 1–3.** The terms in the key below relate to processes in the carbon cycle. Match a letter from the key with each of the following questions. A letter may be used once, more than once, or not at all.

(A) carbon neutrality
(B) carbon sequestration
(C) cellular respiration
(D) carbon flux from atmosphere to lithosphere
(E) fossil fuel combustion

1. The process whereby the biota of ecosystems release energy necessary for functioning

2. The main ecosystem service that permanently protected forests provide

3. The main cause of a lowering of pH in marine environments

**Questions 4–7**
Base your answers to questions 4 through 7 on the following diagram:

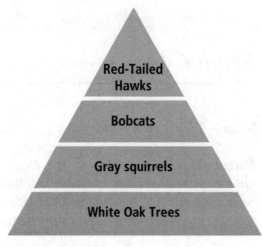

4. The greatest amount of energy present in the above ecosystem is found at the level of
   (A) red-tailed hawks.
   (B) bobcats.
   (C) gray squirrels.
   (D) white oak trees.
   (E) decomposers.

5. In the above ecosystem, 250 kilograms of bobcats would require ___, in order to live and grow.
   (A) 125 kg of gray squirrels
   (B) 250,000 kg of autotrophs
   (C) 2,500 kg of gray squirrels
   (D) less than 2,500 kg of herbivores
   (E) 62.5 kg of white oak trees

6. The specific tertiary consumers in this ecosystem are
   (A) gray squirrels.
   (B) bobcats.
   (C) red-tailed hawks.
   (D) omnivores.
   (E) non-existent.

7. This ecosystem would most likely be found in the biome known as the
   (A) savanna forest.
   (B) boreal forest.
   (C) taiga.
   (D) tropical rainforest.
   (E) temperate deciduous forest.

8. A city with a population of 10,000 is experiencing 20 deaths, 40 births, 10 emigrants, and 20 immigrants per year. What is the net annual percentage growth rate of this city?
   (A) 80%
   (B) 1%
   (C) 3%
   (D) 0.3%
   (E) 0.1%

9. A country that is beginning to go through demographic transition and becoming industrialized first experiences
   (A) death rates rising as birth rates fall.
   (B) birth rates rising.
   (C) death rates rising.
   (D) birth rates falling.
   (E) high death rates until fully industrialized.

10. A scientist does an experiment with *Daphnia*, a sensitive aquatic organism, to determine the potential toxicity of an herbicide. Eighty individuals were used, and the scientist finds that 10 mg/L kills half the *Daphnia*. The scientist has now determined
    (A) the TC-50.
    (B) the ED-50.
    (C) the potential lethality of this pesticide to humans.
    (D) the LD-50.
    (E) a dose-response curve that can be applied to humans.

11. The main chemical *causing* the "dead zone" in the Gulf of Mexico is
    (A) arsenic.
    (B) mercury.
    (C) oxygen.
    (D) phosphorous.
    (E) nitrogen.

12. The main chemical responsible for acid deposition is
    (A) carbon dioxide.
    (B) sulfur dioxide.
    (C) CFCs.
    (D) mercury.
    (E) smog.

13. In terms of ecosystem services, a sustainably managed ecosystem would
    (A) restore the original biodiversity of the ecosystem.
    (B) be more expensive to initiate, but overall healthier in terms of biodiversity.
    (C) provide more valuable services than converted ecosystems.
    (D) provide enhanced and restored services over unaltered natural ecosystems.
    (E) mitigate most air and water pollution problems.

14. The most productive (g C/m$^2$/yr) ecosystems in the world are
    (A) temperate deciduous forests.
    (B) swamps and marshes.
    (C) the open ocean.
    (D) temperate grasslands.
    (E) northern coniferous forests.

15. If a coral reef has an NPP of 5.25 kg C/m$^2$/yr and a GPP of 7.5 kg C/m$^2$/year, how much carbon is being used during respiration by the autotrophs in this ecosystem?
    (A) 12.75 kg C/m$^2$/year
    (B) 2.25 kg C/m$^2$/year
    (C) −2.25 kg C/m$^2$/year
    (D) 14.95 kg C/m$^2$/year
    (E) cannot be determined with the information given

**Questions 16–18**

Use the diagram below to answer questions 16 through 18.

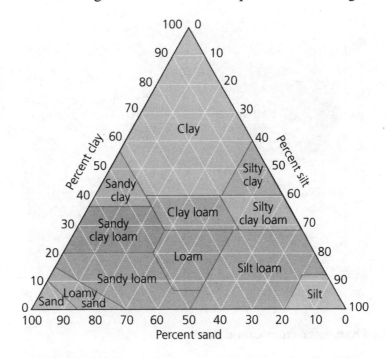

16. If a soil sample was determined to be composed of 40% sand, 20% silt, and 40% clay, the soil type would most closely resemble
    (A) sandy clay loam.
    (B) silt loam.
    (C) silty clay loam.
    (D) sandy clay.
    (E) silty clay.

17. The soil sample that would have the *largest runoff* after a rainfall event would be
    (A) silty clay.
    (B) sandy clay.
    (C) silt loam.
    (D) sandy loam.
    (E) sand.

18. The best soil type for most agriculture would be composed of
    (A) 80% sand, 40% silt, 10% clay.
    (B) 40% sand, 40% silt, 20% clay.
    (C) 20% sand, 60% silt, 20% clay.
    (D) 35% sand, 35% silt, 30% clay.
    (E) 33.3% sand, 33.3% silt, 33.3% clay.

**Questions 19–21**

Use the graph below to answer questions 19 through 21.

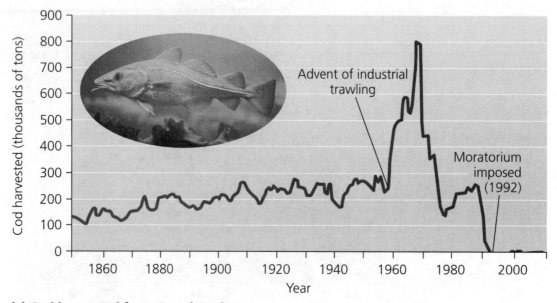

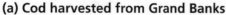
**(a) Cod harvested from Grand Banks**

19. The change in the cod harvested from the Grand Banks between 1960 and 1980 has been a
    (A) 200% decline.
    (B) 100% decline.
    (C) 40% decline.
    (D) 20% decline.
    (E) 10% decline.

20. This initial decline has been attributed primarily to
    (A) foreign fishing fleets in Canadian waters.
    (B) international fishing fleets using bottom-trawling nets.
    (C) large by-catch harvesting.
    (D) drift-net fishing by Japanese tuna fisherman.
    (E) inadequate population estimates made by the Canadian government.

21. The current rate of cod harvesting is primarily explained by
    (A) the ability of cod to rebound after the moratorium.
    (B) the overharvesting of by-catch.
    (C) the overharvesting of keystone species on the Grand Banks.
    (D) the inability of all to agree on the requirements of the moratorium.
    (E) the collapse of the cod population on the Grand Banks.

**Questions 22–25.** Match a letter from the following list of energy sources with each of the following questions. A letter may be used once, more than once, or not at all.

(A) Biofuels
(B) Wind
(C) Shale deposits
(D) Hydropower
(E) Sunlight

22. The source that is converted directly into electricity by PV cells

23. The source whose use is potentially in direct competition with the global food supply

24. Hydraulic fracturing is the methodology used to harvest the energy reserves from this source

25. In terms of efficiency, this energy source has the highest EROI
    **Questions 26–28** refer to the following graph:

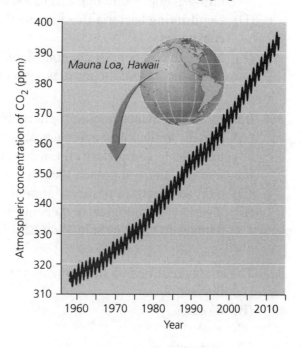

26. The annual fluctuation of the $CO_2$ concentration is explained by
    (A) annual absorption of $CO_2$ by the oceans.
    (B) the ozone layer mitigating the annual increase in $CO_2$.
    (C) seasonal photosynthetic activity of autotrophs.
    (D) annual El Niño events.
    (E) annual respiration rates by global biota.

27. The data in the graph can be evidence for global climate change when they are combined with
    (A) fossil fuel consumption by the United States.
    (B) the global ecological footprint.
    (C) annual cycles of flooding and drought.
    (D) annual CFC production.
    (E) mean global temperatures.

28. The % change in $CO_2$ concentration between 1965 and 2013 is approximately
    (A) 80%.
    (B) 70%.
    (C) 35%.
    (D) 25%.
    (E) 10%.

29. Which of the following is true of nitrogen as it cycles in nature?
    (A) The largest sink of nitrogen is in the atmosphere.
    (B) Nitrogen compounds rarely exist in the gaseous state.
    (C) Nitrogen fixation occurs regularly in ecosystems, producing a nitrogen compound that can be used by primary consumers.
    (D) Denitrifying bacteria live in mutualistic relationships with most consumers.
    (E) Nitrogen compounds are considered the least problematic to the environment and the most essential to all animal life on Earth.

30. Integrated pest management programs are *primarily*
    (A) organic agricultural approaches.
    (B) methods to completely eliminate agricultural pests.
    (C) a monoculture approach to agriculture.
    (D) a broad-spectrum pesticide approach to eliminating pests.
    (E) a variety of biological and chemical approaches to reduce crop damage to an economically tolerable level.

31. The energy returned on investment, or EROI value, is the lowest for
    (A) nuclear power.
    (B) biofuels.
    (C) coal.
    (D) hydropower.
    (E) PV technology.

32. Which of the following greenhouse gases has the *highest* heat-trapping ability per molecule in our atmosphere?
    (A) NO
    (B) $CH_4$
    (C) $CO_2$
    (D) CO
    (E) $SO_2$

33. Assuming that the human population of Earth is growing at an annual rate of 1.2%, how many years will it take for the world population to double?
    (A) 2.4 years
    (B) 4.8 years
    (C) 58 years
    (D) 100 years
    (E) 140 years

34. When comparing nuclear and coal-fired power plants, a nuclear power plant generates more
    (A) $NO_2$.
    (B) radon.
    (C) $CO_2$.
    (D) thermal pollution.
    (E) fly ash.

35. Characteristics that contribute to the vulnerability of a species becoming endangered include which of the following?
    I.    having low genetic diversity
    II.   having a generalist habitat
    III.  being a keystone species
    IV.   being r-selected

    (A) I only
    (B) II only
    (C) III only
    (D) III and IV only
    (E) I and IV only

36. The international agreement that is widely considered the most important environmental victory for a sustainable planet is
    (A) the Kyoto Protocol.
    (B) the Montreal Protocol.
    (C) the Clean Air Act of 1972.
    (D) NAFTA.
    (E) the Green Revolution.

37. The primary characteristics of sewage treatment plant effluent are
    (A) moderate oxygen, low coliform, low BOD.
    (B) high oxygen, high coliform, low BOD.
    (C) low oxygen, low coliform, high BOD.
    (D) high nitrates, high phosphates, low BOD.
    (E) low nitrates, low phosphates, high BOD.

38. When the I = PAT model is applied to the United States, the *most* important factor in determining the environmental effects of the U.S. population is
    (A) the value of P.
    (B) the value of A.
    (C) the value of T.
    (D) the value of P and T.
    (E) the value of TFR.

**Questions 39–41**
Answer questions 39 through 41 by choosing the species that *best* represents the quality asked for.

        I.    amphibians
        II.   bald eagles
        III.  sea otters
        IV.  giant pandas

39. Most species have particular roles or characteristics that make them important and unique in their native ecosystem. Species that are the *most* affected by biomagnification issues is/are
    (A) I only.
    (B) II only.
    (C) III only.
    (D) I, II, and III.
    (E) III and IV only.

40. The *best* example(s) of an indicator species is/are
    (A) I only.
    (B) II only.
    (C) III only.
    (D) IV only.
    (E) III and IV only.

41. The *best* representation of keystone species is/are
    (A) I only.
    (B) II only.
    (C) III only.
    (D) I, II, and III.
    (E) III and IV only.

42. All of the following are examples of GMOs *except*
    (A)  golden rice.
    (B)  Bt corn.
    (C)  Roundup-Ready alfalfa.
    (D)  Monsanto meat.
    (E)  Bt canola oil.

**Questions 43–44**
Use the graph below to answer questions 43 and 44.

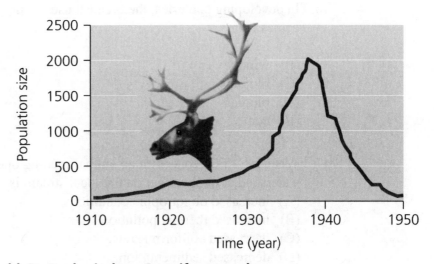

**(c) St. Paul reindeer, *Rangifer tarandus***

43. The reindeer's population growth between the years 1920 and 1940 *best* exemplifies
    (A)  logistic growth.
    (B)  demographic transition.
    (C)  doubling time.
    (D)  K-selection.
    (E)  logarithmic growth.

44. When examining the entire period between 1910 and 1950, the data *most likely* show that
    (A)  the reindeer's environment provided unlimited resources.
    (B)  there were other species competing for food resources.
    (C)  introduced species arrived in 1930.
    (D)  the reindeer's carrying capacity was reached around 1938.
    (E)  limiting factors are not influencing reindeer from 1940–1950.

45. A classroom uses eight 60-watt lightbulbs for 10 hours per day. Approximately how many kilowatt-hours of electricity are used per day by the classroom's lightbulbs?
    (A) 24
    (B) 48
    (C) 480
    (D) 600
    (E) 4,800

46. In developing countries, the greatest use of energy for domestic use comes from
    (A) hydropower.
    (B) windpower.
    (C) biofuels.
    (D) biomass.
    (E) coal.

47. When a CAFO (concentrated animal feeding operation) is located in a watershed, a likely effect on the local stream is
    (A) increased oligotrophic conditions.
    (B) increased thermal pollution.
    (C) decreased coliform levels.
    (D) decreased sedimentation.
    (E) increased eutrophic conditions.

48. The heavy metal produced when coal is burned is
    (A) mercury.
    (B) radon.
    (C) lead.
    (D) iron.
    (E) cadmium.

49. The pollutant now commonly detected in many domestic water wells surrounding hydraulic fracturing operations is
    (A) lead.
    (B) methane.
    (C) natural gas.
    (D) mercury.
    (E) atrazine.

50. The fossil fuel that is in the largest supply globally is
   (A) oil.
   (B) natural gas.
   (C) coal.
   (D) biomass.
   (E) biofuels.

51. Cultural eutrophication is caused primarily by
   (A) nitrogen.
   (B) phosphorous.
   (C) calcium.
   (D) carbon.
   (E) sulfur.

52. The growing concern surrounding ocean acidification focuses on the main causative agent of
   (A) decreasing calcium in oceans.
   (B) increasing pH in oceans.
   (C) increasing $CO_2$ in oceans.
   (D) global overfishing.
   (E) decreasing carbonic acid in oceans.

**Questions 53–55**
Use the following graph to answer questions 53 through 55.

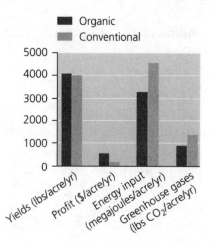

53. A major conclusion that can be made from the evidence comparing organic and conventional farming is
    (A) conventional farming has a proven history.
    (B) organic farming produces more greenhouse gases.
    (C) conventional farming is more economic yet more environmentally problematic.
    (D) organic farming methods should be more widely considered because of numerous, proven benefits.
    (E) organic farming methods are more challenging to institute; however, energy inputs are greater.

54. Environmentally, the *most* significant data provided by the graph, demonstrating the disadvantages of conventional farming, are in the category of
    (A) yields.
    (B) profit.
    (C) energy input.
    (D) greenhouse gases.
    (E) internal costs.

55. A *primary* reason for the differences in energy inputs between conventional and organic farming is that
    (A) IPM approaches and animal manure are being regularly employed by organic farms.
    (B) no-till farming practices are the main energy-saving practices used by conventional farms.
    (C) pesticides, herbicides, and inorganic fertilizers are used widely in both conventional and organic farming approaches.
    (D) organic farming routinely uses animals rather than tractors.
    (E) conventional farming saves energy by employing monoculture approaches.

**Questions 56–58:** The terms in the key below refer to important environmental legislation. Match a letter from the key with each of the following questions. A letter may be used once, more than once, or not at all.

(A)  Clean Water Act
(B)  Resource Conservation and Recovery Act (RCRA)
(C)  Comprehensive Environmental Response Compensation and Liability Act (CERCLA)
(D)  National Environmental Policy Act (NEPA)
(E)  Safe Drinking Water Act

56.  Requires cradle-to-grave tracking of hazardous wastes

57.  Requires an environmental impact statement and study prior to federal projects that might affect environmental quality

58.  Establishes basic coliform levels accepted for human health

59.  An example of a *secondary* extraction process is
(A)  mountaintop removal for coal.
(B)  ocean drilling for oil.
(C)  mining for uranium.
(D)  combustion of biomass waste products.
(E)  hydraulic fracturing of shale.

60.  A cap-and-trade system
(A)  specifies a certain limit on industrial pollutants that can be negotiated with other nations.
(B)  allows nations to negotiate whether or not they will contribute to pollution.
(C)  permits industries that pollute under federal limits to sell credits to industries that pollute over federal limits.
(D)  has recently been repealed by Congress as useless in diminishing pollution.
(E)  allows industries to set their own levels for pollution so that trading can continue.

61.  Environmentally, the main criticism of cost-benefit analyses for public projects is that they *rarely* include
(A)  the internal costs.
(B)  the external costs.
(C)  the profits involved.
(D)  all benefits involved.
(E)  the GDP.

62. A "tragedy of the commons" scenario is *best* exemplified by
    (A) the U.S. agribusiness in the Midwest.
    (B) the extirpation of cheetahs in Africa.
    (C) the extinction of the Great Auk.
    (D) overgrazing sheep in a private park.
    (E) acid precipitation falling throughout the northeastern United States.

63. Structures given a LEED platinum ranking are
    (A) constructed sustainably without electricity from fossil fuels.
    (B) free of asbestos, lead, and radon and therefore made of "green" construction materials.
    (C) especially high in heavy metal content.
    (D) inefficient and need to be renovated in order to meet "green" standards.
    (E) exemplary in sustainable design, especially energy efficiency.

64. A trophic cascade is best described as the effect of ___ on ___.
    (A) top predators; top carnivores
    (B) top predators; abundance of lower consumers
    (C) herbivores; carnivores
    (D) autotrophs; herbivores
    (E) top carnivores; one another

65. Global climate change could be best thought of as a(n) ___ affecting a population's size.
    (A) logistical growth factor
    (B) endemic limiting factor
    (C) density-dependent factor
    (D) density-independent factor
    (E) exponential growth factor

66. The largest reservoir of carbon in the carbon cycle is
    (A) the oceans.
    (B) freshwater systems.
    (C) plant life.
    (D) sedimentary rock and fossil fuels.
    (E) located in the atmosphere.

67. If the global population has a CDR of 11 and a CBR of 17, this would result in a growth rate of ___%.
    (A) 0.06
    (B) 0.6
    (C) 6.0
    (D) 28
    (E) 60

68. Forests supply us with many vital ecosystem services. The one receiving the most attention currently because of global climate change is
(A) the photosynthetic contribution of oxygen from the world's forests.
(B) cap-and-trade potential for global forests.
(C) the influence of El Niño effects by global forest distribution.
(D) moderation of precipitation and climate patterns.
(E) carbon sequestration potential of the world's forests.

69. Sustainable forest certification (SFC) involves
(A) making significant conservation concessions.
(B) a chain of custody following for all harvested materials.
(C) protecting all primary growth forests.
(D) a shelterwood approach to forest management.
(E) controlled burns on a regular basis.

70. From an environmental waste perspective, which approach is the *most* effective and desirable for a sustainable future?
(A) Compost
(B) Reduce
(C) Reuse
(D) Recycle
(E) Incinerate

71. Bisphenol A (BPA) is a chemical used in ___ and has the effect of ___.
(A) pesticides; a carcinogen
(B) herbicides; a neurotoxin
(C) plastics; an endocrine disruptor
(D) food additives; a teratogen
(E) synthetic fuels: a carcinogen

72. Which of the following is *not* an environmental benefit of smart growth?
(A) decreased open space
(B) decreased impervious surfaces
(C) decreased fossil fuel consumption
(D) decreased water pollution
(E) increased use of urban growth boundaries

73. Humans use fresh water primarily for
(A) electrical production.
(B) mining and industrial processes.
(C) agricultural irrigation.
(D) residential heating, cooking, and plumbing.
(E) global nuclear power energy production.

74. Which of the following is not yet regulated by the U.S. Environmental Protection Agency?
    (A) CO
    (B) $O_3$
    (C) $CO_2$
    (D) particulates
    (E) Pb

75. The greenhouse effect involves warming Earth's surface and the
    (A) ionosphere.
    (B) thermosphere.
    (C) mesosphere.
    (D) troposphere.
    (E) stratosphere.

76. "Peak oil" is an environmental topic referring to
    (A) the amount of oil reserves remaining as compared to nuclear fuels.
    (B) the amount of oil reserves that have already been consumed.
    (C) the point at which world oil production will reach a maximum, coinciding with the point at which we will run out of oil.
    (D) the point at which half the total known oil supply is used up.
    (E) the point at which the "technically recoverable" coincides with the "economically recoverable" oil reserve estimates.

77. The Marcellus Shale deposits of Pennsylvania represent known energy reserves of
    (A) oil.
    (B) coal.
    (C) tar sands.
    (D) natural gas.
    (E) biofuels.

78. One's ecological footprint can best be defined as
    (A) the environmental impact on global air and water.
    (B) the effects of the current human population.
    (C) the total amount of waste generated from the human population.
    (D) the area of land and water required to support needs and to absorb the wastes produced.
    (E) the approach of leaving only footprints when visiting national parks, forests, and wildlife refuges.

79. Given that a specific radioactive isotope has a half-life of 30 days, how many days will it take for a 16 kg sample to decay to less than 1 kg?
    (A) 60 days
    (B) 120 days
    (C) 150 days
    (D) 15 days
    (E) 7.5 days

80. Quagga and zebra mussels are best known as
    (A) endangered species.
    (B) keystone species.
    (C) umbrella species.
    (D) invasive species.
    (E) indicator species.

81. As of 2017, the world human population is approximately
    (A) 6.5 billion.
    (B) 7.4 billion.
    (C) 10.3 billion.
    (D) 300 million.
    (E) 20 billion.

82. Transgenic organisms are primarily
    (A) endemic species.
    (B) species that live in border regions between two countries.
    (C) extirpated species.
    (D) genetic species that are protected.
    (E) genetically engineered organisms.

83. The world's leader in producing and using the most solar power is currently
    (A) the United States.
    (B) France.
    (C) Germany.
    (D) the U.K.
    (E) Japan

84. The nation obtaining the highest percentage of its electricity from nuclear power is
    (A) Sweden.
    (B) France.
    (C) Germany.
    (D) Russia.
    (E) Japan.

85. Paleoclimatology data used to study global climate change can be obtained from
    (A) lichens.
    (B) redwood trees.
    (C) amphibians.
    (D) Antarctic ice cores.
    (E) Greenland glaciers.

86. Current sea-level rise has been estimated to be closest to
    (A) 0.5 mm/yr.
    (B) 1.0 mm/yr.
    (C) 3.0 mm/yr.
    (D) 10 mm/yr.
    (E) 20 mm/yr.

87. Biodiversity "hotspots" are
    (A) global areas currently experiencing the most global climate change.
    (B) sites in the U.S. where hazardous wastes have been located.
    (C) priority areas for habitat preservation because they contain many endemic species.
    (D) tropical rainforest areas where there is the greatest species richness.
    (E) areas containing the most endangered species.

**Questions 88–90**
Use the following to answer questions 88 through 90.

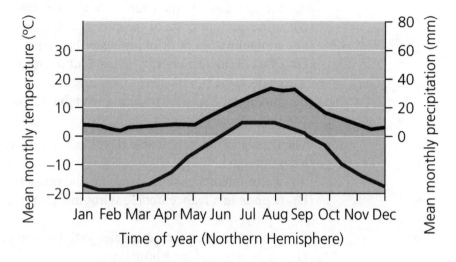

88. The above climatograph shows data characteristic of the ___ biome.
    (A) desert
    (B) savanna
    (C) tundra
    (D) temperate deciduous
    (E) taiga

89. The average annual temperature in this biome is
    (A) above freezing.
    (B) below freezing.
    (C) at freezing.
    (D) cannot be determined.
    (E) approximately 5° C.

90. One of the most distinctive characteristics of this biome is the presence of
    (A) coniferous trees.
    (B) saguaro cactus
    (C) rocky/sandy terrain.
    (D) permafrost.
    (E) rattlesnakes.

91. Campus sustainability efforts are diverse in the United States; however, by far the most frequently adapted have been those using
    (A) renewable energy generation.
    (B) LEED certification.
    (C) discounted mass transit.
    (D) water efficiency upgrades.
    (E) recycling and composting.

92. The *best* example of an external cost resulting from hydraulic fracturing energy projects is
    (A) the land costs to the energy companies involved.
    (B) infrastructure costs of the energy extraction processes.
    (C) contamination of drinking water by methane.
    (D) costs of nuclear energy investment for initially establishing the project.
    (E) endangered species resulting from habitat destruction.

93. When considering sustainable development, the triple bottom line refers to combining
    (A) environmental, social standing, and economic goals.
    (B) economic advancement, environmental protection, and social equity.
    (C) protecting endangered species, habitat, and genetic diversity.
    (D) air, water, and land protection.
    (E) legislation, sustainability, and LEED structures.

94. Deepwater Horizon and Exxon Valdez refer to
    (A) U.S. energy exploration projects.
    (B) the international habitat protection initiative.
    (C) marine oil spills in U.S. waters.
    (D) Alaskan oil drilling projects.
    (E) Gulf of Mexico marine habitat protection projects.

**Questions 95–97**
Use the Age Distribution diagram below to answer questions 95 through 97.

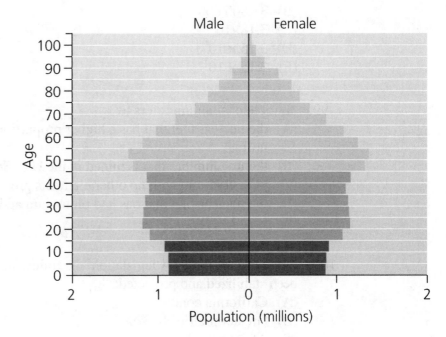

95. Country "X" has the above age distribution. These data indicate that country "X"
   (A) is a developing country.
   (B) has 80% of its population over age 50.
   (C) is experiencing rapid growth.
   (D) is experiencing slow or no population growth.
   (E) is experiencing a high crude birth rate.

96. Country "X" is most likely
   (A) China
   (B) Mexico
   (C) Nigeria
   (D) Kenya
   (E) Canada

97. From these data, this country's growth rate is most likely
   (A) 0.4%.
   (B) 1.0%.
   (C) 3.0%.
   (D) 6.0%.
   (E) 10%.

98. If a country's population is 315 million in January and is growing at a rate of 1%, its population will be ___ at the end of December.
    (A) 3,150,000
    (B) 31,500,000
    (C) $3.18 \times 10^8$
    (D) $3.15 \times 10^8$
    (E) $3.15 \times 10^{10}$

99. "Enriched" uranium refers to
    (A) the nuclear fuel that has a higher proportion of U-235 nuclei than normal.
    (B) the uranium ore that is mined in the U.S. Southwest.
    (C) pure uranium, with no other elements present.
    (D) the nuclear fuel that has had plutonium added to it.
    (E) pure U-238.

100. As a result of the ESA legislation, the following endangered species has been stabilized and protected:
    (A) California condor
    (B) Greater prairie chicken
    (C) African elephant
    (D) Black rhino
    (E) Costa Rican golden toad

# Environmental Science

## *Section II*

**Time—90 minutes**
**4 Questions**

**Directions:** Answer all four questions, which are weighted equally. It is suggested that each question take approximately 22 minutes to answer. Write all answers in essay form. Outline is NOT acceptable. Where calculations are required, clearly show how you arrived at your answer. Where explanations are required, support your answers with relevant information and/or specific examples. It is important that you read each question completely before you begin to write.

**No calculators may be used in this section of the examination.**

1. Read the *Freemont Press* article below and answer the questions that follow.

---

**FREEMONT PRESS**          January 2, 2017

### *Town Council Debates Coastal Aquaculture Facility*

Last night the Freemont Town Council met to hear the results of the past year's scientific studies, which examined the environmental effects and economic need for the Mighty Mussels Shellfish Aquaculture plant. This local facility began operating in 2014. Advocates have spoken of the economic and health benefits, while critics have voiced concerns about possible negative environmental effects. The recent scientific studies focused on the increasing concerns surrounding ocean acidification. A vote is scheduled for later this month to decide whether to allow additional aquaculture facilities to locate within the city limits.

---

   a) Discuss/explain *why* the issue of ocean acidification would become a central question surrounding the vote to support a shellfish aquaculture facility.
   b) Using seawater chemistry, explain the chemical reactions involved in ocean acidification.
   c) Suggest a viable energy alternative for the coastal community of Freemont, explaining why your option would promote a sustainable energy future and would *not* contribute to further global ocean acidification.
   d) Explain two specific benefits and two possible negative impacts of aquaculture facilities.

2.  The Jones family currently uses an average of 750 kilowatts for their home electricity (hot water, lights, all electrical appliances). Their local utility is charging $0.12 per kWh. After doing research about the new PV solar technology that is increasing in popularity in their state, the Jones family decided to install PV panels on their roof. These panels will produce an average of 450 kilowatts per month, and will have a total cost of $3,500 to install.

    a)  How much did the Jones family pay for their electricity per month *before* the installation of their solar panels?

    b)  How many years will it take for the Jones to break even on the purchase of their solar panels?

    c)  Describe two passive solar energy approaches that the Jones family can implement that would further reduce their electricity consumption.

    d)  The Jones family's utility company is currently using natural gas obtained from hydraulic fracturing taking place in their state. Describe two environmental concerns connected with this approach to obtain energy.

3.  As ecologists know, understanding ecosystem structure and function is the basis of understanding and solving all environmental problems.

    a)  <u>Diagram</u> a specific *temperate deciduous forest biome* with 4 trophic levels, including 1 specific species per trophic level.

    b)  Assume that the biomass of the organisms at the 1st trophic level is 6,500 kg. Calculate and diagram the remaining biomass in all other trophic levels, providing your explanation for the values you determine.

    c)  Assume that an area adjacent to this forest has been plowed and has been growing food crops with the use of modern agricultural approaches, including the use of fertilizers and pesticides. Which trophic level and specific organisms in your temperate deciduous forest biome would be most affected by the nearby farming practices? Explain.

    d)  Trace the path of nitrogen from the atmosphere into the 1st trophic level, explaining how this occurs and showing 3 different *specific chemical forms* which nitrogen takes during the process of assimilation into plant protein.

**4.** Seven U.S. southwestern states and Mexico remove water from the Colorado River for a variety of purposes. Central California has annually removed more than its fair share for agricultural purposes, although *all uses* are supposed to be regulated by the Colorado River Compact of 1922. As a result of the competing interests of southwestern U.S. states and Mexico, currently only a trickle of the Colorado River reaches the Gulf of Mexico. In 2015, in response to a growing concern over an already four-year drought, Governor Jerry Brown announced a 25% mandatory water reduction for all California cities and towns.

a) Explain how current Colorado River usage is a "tragedy of the commons."

b) In response to the above issue, describe/explain *three specific and feasible* approaches that a California citizen could take to live more sustainably.

c) Describe/explain *two* specific ecosystem services connected with the Colorado River.

d) For *one* of your ecosystem services described, explain the specific effects *on the natural environment* of this now compromised ecosystem service.

e) Other than its effect on California's agricultural sector, describe two additional environmental consequences of California's current historic drought.

# Practice Test Answer Key

*<u>Section I</u>: Multiple Choice*

| | | | | |
|---|---|---|---|---|
| 1. C | 25. D | 49. B | 73. C | 97. A |
| 2. B | 26. C | 50. C | 74. C | 98. C |
| 3. E | 27. E | 51. A | 75. D | 99. A |
| 4. D | 28. D | 52. C | 76. D | 100. A |
| 5. C | 29. A | 53. D | 77. D | |
| 6. C | 30. E | 54. C | 78. D | |
| 7. E | 31. B | 55. A | 79. C | |
| 8. D | 32. A | 56. B | 80. D | |
| 9. D | 33. C | 57. D | 81. B | |
| 10. D | 34. D | 58. E | 82. E | |
| 11. E | 35. E | 59. E | 83. C | |
| 12. B | 36. B | 60. C | 84. B | |
| 13. C | 37. A | 61. B | 85. D | |
| 14. B | 38. B | 62. E | 86. C | |
| 15. B | 39. B | 63. E | 87. C | |
| 16. D | 40. A | 64. B | 88. C | |
| 17. A | 41. C | 65. D | 89. B | |
| 18. B | 42. D | 66. D | 90. D | |
| 19. C | 43. E | 67. B | 91. E | |
| 20. B | 44. D | 68. E | 92. C | |
| 21. E | 45. E | 69. B | 93. B | |
| 22. E | 46. D | 70. B | 94. C | |
| 23. A | 47. E | 71. C | 95. D | |
| 24. C | 48. A | 72. A | 96. E | |

### Section II: Explanations for Free-Response Questions

Each Free-Response Question has the potential of earning a total of 10 points. The following point allocations are those suggested for each question part. For some question parts where more than one answer is required, the list of possible responses provided include the *main/central concepts*; however, they are not necessarily exhaustive for all possibilities.

**FRQ 1. a)** Ocean acidification, the process by which the pH of the ocean waters is lowering (becoming more acidic), would be of interest to a shellfish aquaculture facility because the pH of the oceans directly determines the availability of carbonate ions that shellfish need in order to build healthy shells. **(1 point)**

**b)** Ocean acidification results from a series of chemical reactions, beginning with excess carbon dioxide from the atmosphere being absorbed by the ocean and forming carbonic acid:

$$CO_2 + H_2O \rightarrow H_2CO_3$$

This carbonic acid dissociates into a bicarbonate ion and a hydrogen ion:

$$H_2CO_3 \rightarrow HCO_3^- + H^+$$

The hydrogen ion then combines with a carbonate ion ($CO_3^{2-}$) to form a second bicarbonate ion:

$$H^+ + CO_3^{2-} \rightarrow HCO_3^-$$

As the availability of carbonate ions in the water declines, it becomes more difficult to build calcium carbonate shells by the process known as calcification. In fact, as carbonic acid and bicarbonate ions become more common in ocean water, the calcium carbonate shells of marine creatures begin dissolving. **(3 points)**

**c)** Since ocean acidification is caused by increasing levels of carbon dioxide in the atmosphere, an energy alternative that does *not* produce carbon dioxide is necessary. Nuclear power does not use a fossil fuel as its energy source; however, there are other environmental problems (primarily the spent nuclear waste disposal issue) to consider. Since Freemont is a coastal town, wind energy would be the most available and environmentally sound. Using the wind as "fuel," with no fossil fuel combustion necessary, Fremont would not be contributing to further global ocean acidification. **(2 points)**

**d)** Benefits of aquaculture include the following: 1) helps to ensure a reliable protein source, 2) can help to reduce fishing pressure on current shellfish stocks, and 3) can be energy-efficient, producing more fish per unit area than is harvested from continental shelf waters or the open ocean, especially since fossil fuels are not needed for powering fishing vessels. Negative impacts of aquaculture often result from the dense concentration of farmed animals resulting in: 1) increased disease necessitating antibiotic treatment and expense, 2) increased animal wastes in water, leading to possible eutrophic conditions, and 3) accidental escape of genetically modified (GMO) fish/shellfish into ecosystems where they could outcompete native species, interbreed with them, or spread disease. (**2 pts. benefits + 2 pts. negative impacts = 4 points**)

**FRQ 2. a)** *Before* installation of the PV panels, the Jones family paid: (**2 points**: 1pt. for work, 1pt. for correct answer):

750 kilowatts/month × $0.12/kilowatt = $90/mo

**b)** The time it will take the Joneses to break even on the purchase of their solar panels (**2 points**):

$$450 \text{ kilowatts/mo} \times \$.12/\text{kilowatt} = \$54/\text{mo}$$

$$\$3500 \div \$54/\text{mo} = 64.8 \text{ months}$$

$$64.8 \text{ months} \div 12 \text{ months/year} = \textbf{5.4 years}$$

**c)** Two *passive* solar energy approaches to further reduce electricity consumption for a New England location:

Passive energy approaches do not have the ability to create electricity; however, through the use of home orientation and/or structural materials, the desired temperature can be obtained through storing/conserving the sun's energy. For any home in a northern latitude (a New England home) this could be in the form of stone floors serving as an internal mass that can directly absorb sunlight's radiant heat. Conversely, planting deciduous trees in front of south-facing windows can serve as shading/cooling during the summer months, not allowing the direct sunlight into a house. (**4 points:** 2 pts for each explanation)

**d)** The primary environmental concerns associated with hydraulic fracturing of shale deposits for natural gas are: 1) large amounts of water usage, depleting groundwater sources needed for other purposes; 2) methane gas contamination of groundwater used for drinking water or other domestic purposes; 3) the need to dispose of the large amounts of contaminated waste water used in the fracturing/drilling process. (**2 points:** choices provided above)

**FRQ 3. a)** 1 point for accurately including temperate deciduous forest species in all 4 trophic levels, and 1 pt. for accurately including the correct trophic level for each species: an autotroph for the $1^{st}$ TL, herbivore in $2^{nd}$ TL, carnivore in $3^{rd}$ TL, and top carnivore in $4^{th}$ TL. Examples follow. No points are awarded if the diagram is in the form of a food web with interconnecting arrows, as it doesn't allow one to accurately determine the specific trophic level to which a species is located. A pyramid diagram is the most accurate format, with each trophic level labeled. **(2 points)**

**b)** **(1 pt.** awarded for *completely* describing the 10% law; **1 pt.** for accurately calculating the energy available for all trophic levels). Merely naming the 10% law does not earn a point.

According to the 10% law, which describes approximate energy transfer between each succeeding higher trophic level, 90% of the energy available at each TL is consumed by the organism's own metabolic processes/needs and lost as heat energy from the ecosystem. Only 10%, at best, is available to the next higher TL. Therefore:

| $6500\,kg \rightarrow$ | $650\,kg \rightarrow$ | $65\,kg \rightarrow$ | $6.5\,kg$ |
|---|---|---|---|
| (autotrophs) | (herbivores) | (carnivores) | (top carnivores) |
| $1^{st}$ TL | $2^{nd}$ TL | $3^{rd}$ TL | $4^{th}$ TL |

**c)** Fertilizers are directed towards the 1st trophic level's autotrophs (plants); however, pesticides would be directed towards 2nd trophic level's insects (plant-eating), and would have *an increasingly magnified effect the higher one goes in the food chain due to biological magnification.* Pesticides are not broken down by normal microbial decomposition and therefore bioaccumulate in each organism, and ultimately biomagnify in the food chain, increasing in concentration the higher the trophic level. **(3 points)**

**d) (3 points)** Nitrogen enters the food chain by the process of ***nitrogen fixation*** (carried out by mutualistic bacteria living in root nodules of legumes), followed by ***nitrification*** (carried out by nitrifying soil bacteria). These chemical processes are:

$$N_2 \text{----(N-fixation)} \rightarrow NH_3 \text{----(Nitrification)} \rightarrow NO_2 \rightarrow NO_3$$
$$\text{(ammonia)} \qquad\qquad \text{(nitrite)}\ \textbf{(nitrate)}$$

**Nitrate ($NO_3$)** is the form of nitrogen which can be taken up and used by plants as a nutrient.

**FRQ 4. a)** The Colorado River holds water in common for seven U.S. states and two northwestern states of Mexico. If California removes more than its share from the river, it leaves less water for the other users, tempting them to do likewise for their own state's uses and personal gain. This poses threats (depletion, eutrophication, pollution) to the entire riverine natural resource, and thus is a "tragedy of the commons." **(2 points)**

**b)** Three feasible approaches that a California citizen could take to live more sustainably would be: **(3 points**: 1 point each)

*4 approaches described—only 3 required.*
   1) increase household water conservation efforts: shorter showers, install conservation shower heads and low-flush toilets.
   2) plant/restore native plants and vegetation around home, as these arid-adapted plants require much less to no additional watering, other than what normally is received in this biome.
   3) eat more local and organic produce from California, thus reducing transportation costs required to ship produce grown in other parts of the country or Central America.
   4) increase recycling and waste reduction efforts, as manufacture of any new product requires water resources and recycling that material requires much less water.

**c)** Two specific ecosystem services associated with the Colorado River are: **(2 points:** *3 specific services described—only 2 required*)
   1) regulation of temperature, precipitation, and cloud formation.
   2) store and regulate water supplies in watersheds and aquifers, necessary for native biodiversity.
   3) when native vegetation has enough water to flourish, protection for soils against storm, flood, and drought damage occurs.

**d)** By removing large amounts of water from the Colorado River (thus depleting levels held in Colorado River reservoirs), there is now less water available for supporting natural biodiversity and the normal functioning of the water cycle. Since 2000, reservoirs are at half-capacity, and climate modelers predict that global climate change will bring still more drought to this area in the next several decades. **(1 point)**

**e)** California's historic drought has affected many aspects of California's environment. (**2 points:** *4 effects described—only 2 required*)

1) In 2015, the U.S. Forest Service estimated that the drought had led to the death of more than 66 million trees in California alone, many killed when trees weakened by the drought were infested by parasitic beetles.

2) As water levels drop and river temperatures rise, the temperature-sensitive Chinook salmon and smelt are experiencing massive die-offs in northern California. This is disrupting aquatic food chains and depriving other fish and predatory birds of important food sources.

3) Since 2012, low water levels in reservoirs have also reduced hydroelectric power generation by 60%, forcing utilities to turn to other energy sources, decreasing overall air quality because of increased fossil fuel use.

4) Many California grassland ecosystems are slowly turning to deserts, reducing further the already endangered kangaroo rat population. This is producing cascading effects on the kangaroo rat predators such as the endangered San Joaquin fox, coyotes, and many birds of prey.